ELEMENTS OF ANTHROPOLOGY
A Series of Introductions

Ecology, Energetics, and Human Variability

Michael A. Little
State University of
New York at Binghamton

George E. B. Morren, Jr.
Cook College, Rutgers University, The State
University of New Jersey

wcb

WM. C. BROWN COMPANY PUBLISHERS
Dubuque, Iowa

Library of Congress Catalog Card Number: 75-23888

ISBN 0–697–07546–X

Printed in the United States of America

Contents

	Preface	v
1	Introduction	1
2	Fundamentals of Ecology	7
3	Human Life-Support Systems	18
4	Human Variability	29
5	Ecological Case Studies of Human Populations	46
	Glossary	85
	Index	88

For Adrienne and Janet

Preface

Human population ecology is a relatively new area of investigation in anthropology; one that has been developing rapidly within the past twenty years. The scope of the field in anthropology is great enough for ecological studies to have taken many directions. Component anthropological fields with identifiable interests in ecology include: cultural ecology, ethnoscience, medical anthropology, and human population biology. Yet anthropology is an integrated science, and in many cases these fields omit very important aspects of the ecology of human populations. For example, biological anthropologists too often lose sight of the sociocultural patterns of a given society and how these patterns aid survival in frequently hostile environments. Cultural ecologists neglect (or merely attempt to deal with as a constant set of variables) humans as biological organisms. And ethnoscientists study a people's views and knowledge of an ecosystem without testing the ecological validity of how this information is used.

Our view of human ecology as a field of investigation within anthropology is that it should treat human populations as an integral part of all the terrestrial ecosystems they inhabit; that human populations respond to the challenges of survival within these ecosystems by behavioral as well as biological means. In the material that follows we have focused on the ecology of technologically less complex societies, where human perturbation of an ecosystem is less than in technologically more complex societies; the ecosystem is likely to be in an approximate state of energetic and nutrient balance. These are also societies with which students of anthropology are likely to have greater familiarity. The task of studying human populations within an integrated ecological framework is a formidable one, but nevertheless one that must be pursued. We hope that this brief exposition will contribute in a small way to the development of this important field within anthropology.

We offer thanks to our many teachers and colleagues who first formulated or developed some of the ideas presented in this work. In particular we wish to thank Andrew P. Vayda and Neville Dyson-Hudson for their critical comments on segments of the work.

Michael A. Little
George E. B. Morren, Jr.

1 Introduction

BIOLOGY AND ECOLOGY

Many fields of study today have a primary interest in the environment. Biometeorologists are concerned largely with the influence of the atmosphere on living forms; public health scientists delve into the environmental conditions contributing to health and disease in human populations; environmental physiologists study the effects of climate, nutrition, activity, and the like on plants and animals; epidemiologists study the distribution and transmission of disease; and some space scientists are interested in the artificial environments needed to sustain human life during spaceflights. These are but a few of the fields in which the environment is a central focus. In one sense these interests are ecological because environment-organism interactions are being studied; in another sense they are not. This distinction will be apparent after a brief discussion of the concept of ecology.

Textbook definitions of "ecology" vary somewhat, but nearly all would come close to the statement that *ecology is the study of the interrelationships among populations of organisms and their environment.* The use of the term "interrelationships" is important because ecologists usually are interested in processes of interaction and integration among organisms and environments. Moreover, ecologists are not so much concerned with the behavior and biology of individual living forms as they are with populations of organisms. These two points of emphasis alone serve to distinguish an ecological study from a biometeorological or public health study, for instance. The concept of an ecological study has been expanded within the past several decades because of an increase in our understanding of how living systems function. If we view a hierarchy of living systems as Odum has, from genes, to cells, to organs, to organ systems, to organisms, to populations of organisms, and through to communities and ecosystems, it is apparent that the next higher level of organization becomes the environment of the preceding level.[1] Hence organs can be viewed as the environment of cells, and one might, as an example, be concerned with "cellular ecology." This also means that certain characteristics of higher-level units may be regarded as summations of attributes of lower-level units. Hence the distinctive dimensions of a population, such as its age-sex structure and birth rate, are the cumulative outcome of the activities of the individual organisms making up the population. The biological hierarchy is a powerful

1. E. P. Odum, *Fundamentals of Ecology,* 3d ed. (Philadelphia: W. B. Saunders, 1971), pp. 4-6.

conceptual tool for reducing the apparent complexity of ecological processes.

Another useful definition of ecology involves looking at the structure and function of ecological relationships of a particular population rather than of a higher-level unit in the hierarchy. From this perspective, ecology is the study of *"any property that has a measurable direct influence on the fecundity, longevity, speed of development, or spatial position of an organism."*[2] With our implied concern for a *human ecology,* this definition provides guidelines promoting a degree of parsimony so often lacking in studies of man. Thus the same framework can be employed to study the influence of measurable environmental and sociocultural (or behavioral) factors on these essential dimensions of a human population.

Ecology has sometimes been defined loosely as *the study of adaptations,* an adaptation being an adjustment (biological or behavioral) that benefits or favors survival of an organism or population within a given environment.[3] This definition points out the relationship between ecology and evolution—the process of organic change through time. An *ecosystem,* comprised both of a community of different species of organisms (*biotic environment*) and its nonbiotic component (*physical environment*), continually changes through time. When they occur over long periods of time, such changes at least partially result from the evolutionary changes in the populations that make up the living community. Since all organisms within a given ecosystem are interdependent to one degree or another, then evolutionary changes depend on the selective pressures that arise from individual organism and population competition for resources within the ecosystem. Thus a comprehensive knowledge of some of the principal mechanisms of evolution can be gained only by studies conducted within an ecological framework.

Although interests in organism/environment relations can be traced back to antiquity, the basis for modern ecology began in the eighteenth and nineteenth centuries.[4] The naturalists of this period systematically gathered vast stores of information that led to insights into the workings of nature and the changes occurring through time. The compilations of the great Swedish taxonomist Linnaeus (1707-78), the writings on population by Malthus (1766-1834), the contributions of the naturalists Compte de Buffon (1707-88), Lamarck (1744-1829), Darwin (1809-82), and Wallace (1823-1913), all led to a recognition of what we now call ecological interactions.[5]

Today, after the period of descriptive and analytical work of the first half of this century, ecological studies are being conducted with a high degree of sophistication. We are now in what can be identified as a "breakthrough" and "synthesis" period in our understanding of the biosphere. Techniques of mathematical modeling and computer simulation have contributed substantially to our knowledge of the factors that influence ecosystem stability and change, diversity, productivity, and population control. As a result, predictive capabilities have been increased manyfold, and the consequences of introducing new elements or organisms to stable ecosystems can be identified. In addition, ecologists are now work-

2. D. A. Maelzor, "A Discussion of Components of Environments in Ecology," *Journal of Theoretical Ecology* 8 (1965):160.

3. M. Bates, "Ecology and Evolution," in *The Evolution of Life, Its Origin, History, and Future,* vol. 1, *Evolution After Darwin,* ed. Sol Tax (Chicago: University of Chicago Press, 1960), pp. 547-68.

4. W. C. Allee et al, *Principles of Animal Ecology* (Philadelphia: W. B. Saunders, 1949), pp. 13-7.

5. L. Eiseley, *Darwin's Century* (Garden City, New York: Doubleday Anchor Books, 1958), pp. 52-5.

ing more closely with population geneticists, demographers, physiologists, and behavioral scientists. The skills required to study ecosystems, especially those of which human populations are a part, are no longer the exclusive property of one discipline (as scientific disciplines have been historically determined). The time is on us for the emergence of a new science, neither exclusively biological nor social, which is prepared to deal with man as an integral part of nature.

ECOLOGY AND THE HUMAN SCIENCES

Much recent knowledge has resulted from the investigations conducted as a part of the International Biological Program (IBP) that ended in 1974. This ten-year worldwide program had as its primary objective the study of "the biological basis of productivity and human welfare" with emphasis on "organic production of the land, in fresh waters, and in the seas" and "human adaptability to the changing conditions."[6] Many of the IBP ecological studies carried out by scientists focused on ecosystems (*biomes*) identified primarily by vegetational characteristics. Coniferous forests, grasslands, Arctic tundra, deciduous forests, and desert biomes were studied by multidisciplinary teams of scientists with a primary objective of fully understanding the structure and function of these incredibly complex systems.

Human adaptability, another major emphasis of the IBP, began with a single, environmentally oriented frame of reference—that of studying the adaptive mechanisms of human populations living under some form of environmental stress. While human populations are enormously adaptable, this adaptability has limits; attempts to exceed these limits produce a toll in reduced function, in disease, and ultimately in death. As a result of rapid technological change, people are not only seriously disturbing their ecosystems but are also subjecting themselves to a wide variety of new stresses. Through studies in genetics, physiology, demography, and anthropology, the human adaptability research programs attempted to understand the processes of human adaptation from various viewpoints. Such studies were conducted among small populations of Alaskan Eskimos and Aleuts and among South American Indians from the Andes and the Amazon rain forest. By studying these technologically simple groups, we have been able to gain important insights into the character of human interaction with the environment from a biological and cultural point of view. These insights might otherwise have been hidden within the complexity of technologically developed human populations. Greater understanding was gained of how people respond to environmental perturbations and of how people respond to particular environmental stresses, such as disease, energy and food availability, climate, and activity requirements.

Human populations are an integral part of most ecosystems. They play an important role in and have a profound impact on the interrelationships that evolve within such systems. Yet human involvements within ecosystems are complex and as such, are difficult to study. Although few truly integrated studies of human ecology have been conducted successfully to date, the potential returns from such studies hopefully will provide us with the information required to maintain the biosphere's delicate balance.

For these hopes to materialize, biological and behavioral scientists will probably

6. U.S. National Committee for the International Biological Program. *U.S. Participation in the International Biological Program,* no. 6. (Washington, D.C.: National Academy of Sciences, 1974), p. 9.

need to critically examine and adopt some of the methods and conclusions developed by each other; the ecological model, especially in combination with systems theory, has great power in unifying science. It can do this by providing a language through which similar questions can be formulated about all living nature, including man. But the study of man presents peculiar problems arising from the complexity of human society and culture, particularly their linguistic and mental structures. The social, behavioral, and historical sciences have developed some useful tools and insights in grappling with such problems.

Physical anthropologists have had fewer problems adopting ecological concepts and methods than their sociocultural colleagues. It is a historical irony that some of the founders of modern biology cited earlier, particularly Malthus and Darwin, proposed ecological models that were explicitly applicable to man as well as to other organisms. However, developments in the sociocultural sciences kept well within the bounds of the idealistic social thought of the same period. Even Herbert Spencer's flawed attempt to make use of Darwinian concepts of fitness and evolution came to stand for the isolation of social science from developments in biology.

As Vayda and Rappaport have pointed out, the crucial difference between a *cultural ecology* rooted in the social science tradition and a unified *general ecology* capable of studying all of life is a conceptual one.[7] In the social science tradition, culture came to be one of the central concepts, particularly in American anthropology. In other words, culture has been the unit of study. By contrast, biologists have focused on the units of the biological hierarchy discussed earlier. The consequences of this conceptual difference can be exemplified in those approaches to human environmental problems labeled "environmentalism" (or "environmental determinism"), "possibilism," and "cultural ecology."

The environmentalist approach holds that environmental factors dictate cultural forms. This position is exemplified in the twentieth century by Ellsworth Huntington's unrigorous statement of the influence of climate on human physiology and psychology.[8] The possibilist position, more characteristic of historically oriented anthropologists, holds that the environment is viewed as permissive—allowing for the free play of cultural, historical, and psychological factors within fairly broad limits.

Julian Steward stands out in his reaction against possibilism and the cultural historical (and "diffusionist") approach. The methodological base supporting this reaction was Steward's attempt to separate man the organism from man the "superorganism". This involved identifying two different components in culture, the *core* and the *rest*. The cultural core consists of "those features which empirical analysis shows to be most closely involved in the utilization of environment in culturally prescribed ways."[9] In practice, "cultural core" came to be synonymous with the technology of food production, although Steward seemed to have had a somewhat broader range of cultural behavior in mind, and in his own work emphasized the relationship between core features and sociopolitical phenomena (further discussed in chapter 3). This would tend to suggest that Steward was not critically applying a biological or ecological model to the subject matter. For example, although in selected cases he was concerned

7. A. P. Vayda and R. A. Rappaport, "Ecology, Cultural and Noncultural," in *Introduction to Cultural Anthropology*, ed. J. A. Clifton (Boston: Houghton Mifflin Company, 1968), pp. 477-97.

8. E. Huntington, *Mainsprings of Civilization* (New York: John Wiley and Sons, Inc., 1945).

9. J. Steward, *Theory of Culture Change* (Urbana, Ill.: University of Illinois Press, 1955), p. 37.

with factors dealing with the biology and behavior of plants and animals exploited by human groups, he seemed to show a lack of interest in the biology and behavior of people. In fact, Steward appeared to equate biology with genetics.[10] Similarly, minimal or no treatment was given to such environmental factors as disease and competing human populations, or to such human biological factors as reproduction.

Our view is that if a general ecological perspective is adopted, the principal unit of study becomes a *human population,* and population-specific cultural behavior and biology may be studied just as the biology and behavior of other species is studied; that is, in terms of the objective material consequences of specific behavioral and biological factors. It is a basic premise of this book that scientific method does not permit a separation of the cultural-behavioral attributes of human adaptations from the physical-biological. In other words, some of the same kinds of questions can be asked concerning the adaptive consequences of human cultural behavior as can be asked about human biological processes, and that the same framework—essentially an ecological one—can be applied conceptually. This is not the only way to look at cultural or biological phenomena, nor the only way to examine such phenomena readily or usefully. Rather, we are concerned with those cultural and biological responses, factors, processes, and cycles that affect or are directly connected with the survival, reproduction, development, longevity, or spatial position of people. This set of questions rather than the traditional division of scientific labor defines the subject matter.

Ample provision exists within this framework for the study of such cultural factors as ideology, values, motivation, linguistic categories, personality, ritual, and the like, insofar as particular manifestations of these influence the interaction of a human population with its environment or affect the biological characteristics of that human population. The possibilities here can be exemplified by investigations of such phenomena as East Asian Indian sacred cattle, religious ritual in New Guinea, ethnoscientific categories and food taboos in New Guinea, and Montagnais-Naskapi (North America) divination.[11] These cases demonstrate that belief is part of resource-management strategies beneficial in some sense to the human populations possessing those beliefs and practices. These factors, as they relate to specific adaptive strategies of human populations, provide the mode in which individuals learn to participate in their population's strategies.

Ecological studies of cultural behavior take into account much more than technology which is merely one of the most obvious classes of cultural behavior. For instance, the above examples involved investigations of the latent functions of belief patterns; that is, of the objective consequences of behavior and belief that were unknown or imperfectly known to the people involved. Possibly, some of the most vital human behavioral influences on ecological systems involve latent functions. Hence it is appropriate to follow a research strategy in which the material aspects and consequences of a wide range of human behavioral responses are studied first, before attempting to bring out their individual-level cultural dimensions.

10. Vayda and Rappaport, "Ecology," p. 488.

11. M. Harris, "The Cultural Ecology of India's Sacred Cattle," *Current Anthropology* 7 (1966):51-66; R. A. Rappaport, *Pigs for the Ancestors: Ritual in the Ecology of a New Guinea People* (New Haven, Conn.: Yale University Press, 1968); G. E. B. Morren, *Settlement Strategies and Hunting in a New Guinea Society,* Ph.D. diss. (Columbia University, 1974); O. K. Moore, "Divination—A New Perspective," *American Anthropologist* 59 (1957):69-74.

With these considerations in mind, a "reformed" definition of culture states that: the culture of a population is a set of learned responses which are (a) *solutions to environmental problems,* and which are (b) *solutions to problems arising from past solutions to environmental problems.* If the term "adaptation" was substituted for "culture" and "learned" deleted in this definition, it would pass as a reasonable definition encompassing the many possible kinds of solutions to environmental problems available to the members of a population. These would then include behavioral responses, physiological responses, and genetic responses. In addition, this definition points to an extremely important characteristic of adaptation—that all successful responses to environmental problems or stresses create new problems. Thus the optimal solution is to "stay in the game."

For Further Reading

Bates, M. *The Prevalence of People.* New York: Charles Scribner and Sons, 1955. A highly readable yet authoritative popular work on human population ecology.

Bresler, J. B., ed. *Human Ecology: Collected Readings.* Reading, Mass.: Addison-Wesley Publishing Company, 1966. A collection of thirty papers from a variety of scientific sources. Some of the papers are quite technical, but many are classic papers.

Rappaport, R. A. "Nature, Culture, and Ecological Anthropology," in *Man, Culture, and Society,* ed. H. L. Shapiro. New York: Oxford University Press, 1971, pp. 237-67. An introduction to ecological anthropology that emphasizes the behavioral as well as the biological focus of general ecology, and discusses the relationship of the ideologist aspect of culture to ecology.

Vayda, A. P., ed. *Environment and Cultural Behavior: Ecological Studies in Cultural Anthropology.* American Museum Sourcebooks in Anthropology. Garden City, N.Y.: The Natural History Press, 1969. Although these twenty-three papers are of mixed quality, a great deal of material in human ecology is covered in this book.

Vayda, A. P. and Rappaport, R. A. "Ecology, Cultural and Noncultural," in *Introduction to Cultural Anthropology,* ed. J. A. Clifton. Boston: Houghton Mifflin Company, 1968, pp. 477-97. A discussion of the history and future of human ecology studies with comments on the need for a truly integrated field of study.

Bibliography

Baker, P. T. 1962. "The Application of Ecological Theory to Anthropology." *American Anthropologist* 64:15-22.

Bates, M. 1962. "Human Ecology." In *Anthropology Today: Selections* ed. S. Tax. Chicago: University of Chicago Press, pp. 222-35.

Hanna, J. M., Friedman, S. M., and Baker, P. T. 1972. "The Status and Future of the U.S. Human Adaptability Research in the International Biological Program." *Human Biology* 44:381-98.

Harris, M. 1968. *The Rise of Anthropological Theory.* New York: Thomas Y. Crowell Company.

Little, M. A. and Friedman, S. M., eds. 1973. *Man in the Ecosystem: Proceedings of a Conference.* University Park, Pa.: IBP Human Adaptability Coordinating Office.

Steward, J. 1938. "Basin-Plateau Aboriginal Sociopolitical Groups," *Bureau of American Ethnology Bulletin* no. 120, Washington, D.C.: Smithsonian Institution.

2 Fundamentals of Ecology

In some ways the earth is a *closed system*: processes take place within the envelopes of the atmosphere, hydrosphere, lithosphere, and biosphere; and materials, including life materials, are continuously recycled within this system. Viewed in another way the earth is an *open system*: it is a dynamic sphere resulting from continual influxes of energy from the sun in the form of visible, ultraviolet, and other kinds of electromagnetic radiation. Energy drives the system. It facilitates the processes of arrangement, breakdown, and rearrangement of the earth's basic raw materials in inorganic, organic, and living levels of organization. Other sources of energy derive from the gravitational pull of the moon and the rotation of the earth.

The water cycle of the earth is an example of some of these processes. Solar heat energy produces evaporation of water molecules from the land and water as well as transpiration (evaporation) of water from plant surfaces. This water, transported to the atmosphere by convective air currents and larger-scale air movements, forms clouds. Clouds redistribute the water through precipitation, which may infiltrate the soil surface of the land or move along the surface of the land as runoff. This latter process contributes to the weathering of the land surface and also aids in redistribution of minerals and organic matter. In this example, the energy from the sun is converted into many forms, such as the kinetic energy of winds as they transport water molecules aloft and the kinetic energy of water as it moves downhill.

The most important means of solar energy conversion for the maintenance of life is *photosynthesis*. In this process, green plants (*autotrophs*, or self-feeders) are able to use the energy from sunlight along with carbon dioxide and water to produce such organic substances as sugar. The organic substances produced by green plants serve as the principal foods (either directly or indirectly) for most animal forms of life (*heterotrophs*, or other than self-feeders). The process of photosynthesis, then, traps and stores solar light energy in the form of living tissue (chemical energy) usable by higher forms of life as food. One can think of all living forms as storage depots of chemical energy.

Although the transfer of chemical energy as food from one life form to another is an exceedingly complex process, it can be simplified somewhat by an illustration of the *food*, or *trophic*, pyramid (see figure 2.1). At the base of the pyramid are the green plants—the *producers*. Producers provide all of the food or chemical energy for all successively higher-level organisms of the pyramid. First-level *consumers* (herbivores) provide food for second-level consumers (*carnivores*) which provide food in turn for third-level consumers, and so on. The *decomposers*—fungi, bacteria, and other microorganisms—break down the remains of living forms into raw organic and inorganic matter, which is deposited in the soil to be used again by the producers. At successively higher levels of the pyramid, the bulk of chemical energy is lost as heat energy, such that fewer and fewer numbers (or weights) of organisms can exist there than at lower levels. An example from the Arctic might include aquatic plants (producers), aquatic invertebrates (first-level consumers), fish (second-level consumers), seals (third-level consumers), and Eskimos (fourth-level consumers).

Food pyramids, however, serve only as rough models for the transfer of energy from one form of life to another. A more realistic representation of these processes is given by use of the *food web*. This is an interconnected network of usually complex feeding relationships. For example, many carnivores prey upon a variety of herbivores and other carnivores. Humans are a prime example of a life-form that falls within a complex food web. To use the example of the Arctic-dwelling Eskimos again, members of this population feed on wild plants, fish, and seals, but may occasionally fall prey to polar bears. Some of the implications of energy and material flows will be discussed in this chapter under the heading "Ecological Processes."

ECOSYSTEM COMPONENTS

At a simple level, an ecosystem can be divided into two major components—*abiotic* and *biotic*. Elements of the abiotic (nonorganic) component (sometimes called the physical environment) include soils, associated minerals and nutrients and water resources, land forms, and climatic characteristics. Variations of these abiotic elements tend to structure but not wholly determine the characteristics of an ecosystem. Soils, the products of physical and chemical weathering processes, are ultimately derived from underlying bedrock materials, although soils may also be transported by gravity, wind action, and water movement. Soils can also be leached of important minerals by water action. Thus we can see how climate and land form can in-

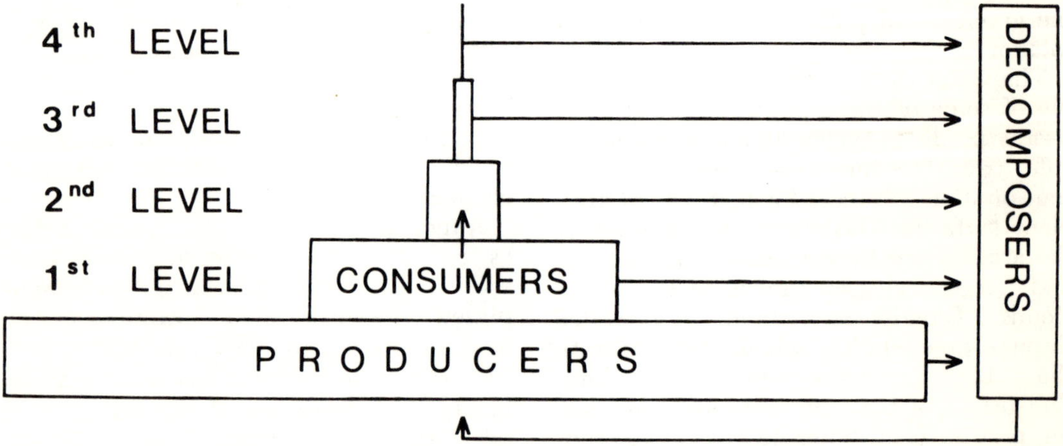

FIGURE 2.1. A representation of a trophic pyramid.

fluence the characteristics of the soils that play such an important role in providing nutrient material for producers. Climatic features over much of the surface of the earth tend to be highly variable. Seasonal fluctuations in solar radiation, temperature, precipitation, evaporation, humidity, cloud cover, and wind flow produce corresponding fluctuations in ecosystems. Temperature and rainfall are perhaps the most crucial variables, or *limiting factors,* in the development of those ecosystems that are highly efficient in trapping and storing energy.

In many ways it is impossible to separate the abiotic from the biotic component of the ecosystem. Soils contains organic matter derived from the biotic components; vegetational cover modifies the effect of climate on weathering processes; and the structure of vegetation even influences climatic characteristics, especially near the surface of the ground. In general though, the biotic component of an ecosystem comprises all forms of life from viruses to humans. Aggregates of organisms of the same species make up *populations,* which tend to exploit the resources of an ecosystem in a given fashion. This pattern of exploitation is identified as a population's *niche* in the ecosystem. A given niche has many limitations based on the abiotic and biotic characteristics of the ecosystem. All of the species or populations within an ecosystem constitute a *community,* characterized by mutual dependency, competition for resources, and complex patterns of interaction among its constituent populations. Ecosystems that are complex and have a high level of *diversity* within the community (such as those found in a tropical forest as opposed to those in a desert) are very stable and tend to persist through time. On the other hand, ecosystems that are made up of only a small number of populations in the community are fragile and easily disrupted by only minor fluctuations in the abiotic or biotic elements of the system.

Thus far we have discussed general principles relating to what are often called "natural ecosystems"; that is, those in which human populations have had minimal or no influence. When human populations are added to biotic communities, they frequently become the dominant life-form because of a superior ability to extract energy and other resources from the ecosystem. Humans nevertheless are an integral part of exploited ecosystems and are subject to the same rules and principles that govern the function of natural ecosystems. One way of viewing human populations within an ecosystem context is presented in figure 2.2. Here, a human-population component of the environment is added to the biotic and abiotic components in a simplified three-component model of ecosystem interaction. The three components interact by way of reciprocal influences, as do elements within each of the three components.

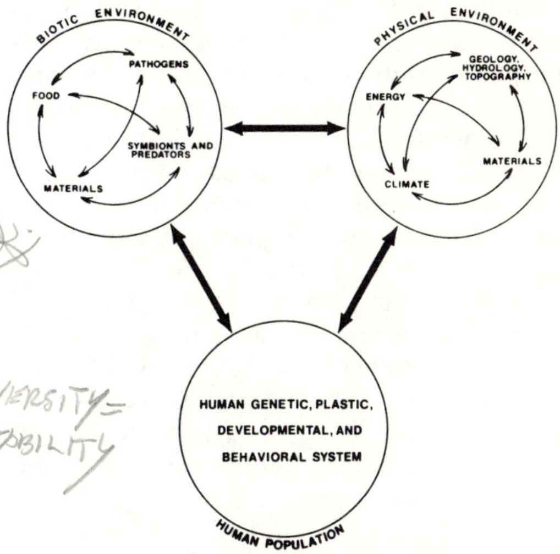

FIGURE 2.2. Major components of an ecosystem of which human populations are a part.

TERRESTRIAL ECOSYSTEMS

Ecosystems can be divided into two major types—terrestrial and aquatic. There are, of course, aquatic elements in nearly all land ecosystems and human populations do exploit aquatic systems. Nevertheless we will focus on the broader aspects of terrestrial ecosystems (communities are called *biomes*) as examples of those inhabited and exploited by man. These are: *tropical forests, temperate deciduous forests, temperate coniferous forests, grasslands, tundras, deserts,* and *mountains.* Each ecosystem is in fact a generalization that shows a wide range of variation as well as a variety of intermediate and intergrading types. Moreover, although some human populations possessing relatively simple technology are a part of a single ecosystem, many other populations extend beyond ecosystem boundaries through networks of trade, communication, and other forms of interchange.

Tropical Forest

The tropical rain forest is characterized by high rainfall and humidity, relatively warm temperatures with greater daily than seasonal variation, and incredible diversity of plant life. Biomes of this type are found in Southeast Asia, New Guinea, the Congo River basin, and the Amazon River basin. Soils tend to fall into the category of *latosols,* thin and relatively poor in nutrient material. Plant productivity and biomass (weight of living matter) here is highest of all the known biomes, but only a relatively small percentage of energy is cycled to consumers because it is trapped as woody support structures (i.e., trees). With the exception of insects, animal diversity and biomass is low.

Temperate Deciduous Forest

Major areas of temperate deciduous forests include much of Europe and the eastern United States. Rainfall is moderately high and constant throughout the year, while temperature, a primary limiting factor to plant growth, shows marked seasonal fluctuations. *Podzol* soils, commonly associated with deciduous forests, are usually thick and rich in organic matter. Productivity of plants is high, but as in the tropical forest, consumer communities make up a relatively small proportion of the total biomass of the biome.

Temperate Coniferous Forest

The evergreen, or boreal, forests of the north are found in a wide belt cross the Soviet Union and throughout much of Canada and the northwest United States. Precipitation is moderate to high while temperature seasonality is characterized by long, cold winters and warm summers. Plant productivity is not only structured and limited by seasonal cold, but also in some areas by thin podzol soils that result in part from slow decomposition of evergreen needles. The northern zones of coniferous forests grade into Arctic tundra.

Grasslands

Grassland biomes can be partitioned into temperate and tropical categories. Typically, both tropical and temperature grasslands are found in areas with moderate to low seasonal rainfall patterns that include periodic droughts. Fire also plays an important role in the maintenance of grassland biomes, especially in moist zones, where in the absence of fire, woodland tends to develop. Temperate grasslands are found in the pampas of Argentina, the steppe of Asia, and the prairie of the United States. Tropical grasslands, or savanna, ecosystems are found in Australia, South America, and Africa. Soils fall into the *chernozem* class, being rich in organic matter and quite thick. Both temperate and savanna grasslands have relatively high plant and animal productivity, with up to 50 percent of the grasses and shrubs being cycled through herbivore and carnivore trophic levels. Herbivore biomass is high in most grassland biomes but reaches a peak in the African savanna, where the diversity of ungulates (hoofed mammals) is very great. Grasslands grade into woodland and forest with increasing moisture or into thorn scrub and desert with decreasing moisture.

Desert

Nearly all continental land masses have deserts—arid zones with rainfall generally less than 25 cm/year or higher rainfall with marked seasonality. Clearly, moisture is the prime limiting factor, although high *evapotranspiration* (evaporation of water from the soil and plant surfaces) increases the aridity in tropical areas. In temperate and tropical desert biomes, daily temperature fluctuations tend to be extreme. Plant productivity is the lowest of all biomes with water-conserving plants predominating; animal life is sparse with low diversity.

Tundra

Tundra lands are distributed along the Arctic Circle of North America, Europe, and Asia. Climatic characteristics are low precipitation, long cold winters and short summers, and minimal solar radiation during the winter months. The subsoil is frozen throughout the year—permafrost—and only the surface layers thaw during the brief growing season. Despite the semiarid conditions, water is abundant during the summer months as a result of its being trapped as ice during the winter and the low evapotranspiration throughout the year. Cold temperatures are most severely limiting to plant growth that reaches a climax of grasses, lichens, and short shrubs during a two- to three-month growing period.

Mountains

Mountain ecosystems are really composites of many other systems, depending on elevation and latitude. Tropical mountain ecosystems begin with rain forest at the base and grade upward through deciduous and coniferous forests, to grasslands with herbs and lichens, and finally to alpine tundra and permanent snow fields. Mountain slopes in a rain shadow will, of course, show characteristics of arid ecosystems. At very high elevations, low temperatures and reduced oxygen are primary limiting factors.

ECOLOGICAL PROCESSES: ENERGY AND MATERIAL FLOWS

Energy is basic to all processes in the universe. All change occurs through transformations of energy from one form to another. Hence in Blum's phrase, energy is "time's arrow," since it makes things happen in the time dimension.[1] Energy and life stand in a special relationship. This is because life is order in a universe characterized by increasing disorder, even though life's orderliness is achieved at a very high energy cost. Many ecologists, including an increasing number of human ecologists, have directed their research toward comprehending energy transformations in living systems at all levels of organization. Here we are concerned with energy as a property of the function of ecosystems, its flow among components of the biotic community, the utilization of energy by populations, and the role of populations and abiotic factors in controlling these flows.

Life = negative entropy

Energy is the ability to do work. The unit of energy employed in this work is a heat equivalent, or kilocalorie (kcal). The same unit is used in nutritional studies and is basically defined as the amount of heat re-

1. H. Blum, *Time's Arrow and Evolution,* 2d ed. rev. (New York: Harper Torchbooks, 1962).

quired to raise the temperature of one kilogram (kg) of water 1°C within a given range. Energy exists in many forms; those of particular significance to biological systems are *mechanical, chemical, radiant* (including light), and *heat*. Basically, there are two forms of mechanical energy—*kinetic* and *potential*. Kinetic energy is free energy associated with a body in motion and can be measured by the work done to bring the body to rest. Examples of kinetic energy include a thrown ball or an automobile in motion. Potential energy is stored energy useful only when it is converted into kinetic energy to perform work. An example is the water stored behind a dam (potential) that is released (kinetic) to produce electricity. Radiant energy typically refers to solar radiation converted into chemical energy by green plants. This chemical energy is in turn the source of energy used by all other living forms.

The Energy Language

The ecologist H. T. Odum developed a graphic shorthand language for representing: (1) the pathways or "food chains" through which energy flows and is transformed in ecosystems, and (2) the control pathways that regulate this process.[2] Odum's graphic language is particularly useful for analyzing the position and role of humans in ecosystems from an energetic standpoint. We have adopted a simplified version of Odum's methods for our work in this chapter and for the case studies that follow.

The basic logic of Odum's method is what he calls his "macroscope," or "detail eliminator." This is a systems approach by which broad features and mechanisms of a system are focused on rather than the details of the parts themselves. The "modules" in the language provide a working list of the energy-transforming functions, and Odum also provides very rough mathematical equations for his modules. Several of Odum's symbols, or modules, are presented in figure 2.3 and described below.

(a) *Source*—A source of energy or some other controlling factor external to the system. These are generally nonliving components of ecosystems including solar radiation, earth and atmospheric conditions, and limiting factors. However, this symbol may also represent such other sources as water from a reservoir, fossil fuel from Saudi Arabia, or food from a store. Sources in effect define the boundary conditions of the system, are located at the "high energy" end of the system, and often have no inputs.

(b) *Passive Storage*—Various forms of stored potential energy. The chemical energy locked in the biomass of a given population, gasoline in a tank, the accumulation of some important element, or grain reserves are examples.

(c) *Heat Sink*—Kinetic energy loss in the form of heat, necessary to all processes involving energy transformations.

(d) *Work Gate*—The work performed by parts of the system in controlling the flow of energy. The flow through the work gate is proportional to the magnitude of the work applied to the top of the gate.

(e) *Self-Maintenance*—The dimension of a population or similar component in a system which is capable of performing work or providing an input to a work gate.

(f) *Green Plant*—Part of the system that is able to receive light energy and convert it into another form of energy (chemical) by photosynthesis.

(g) *Switch*—A form of work gate possessing only "on" and "off" states.

(h) *Transaction*—Economic exchanges involving goods or money and energy flows. It is an "economic" work gate.

(i) *Constant Gain Amplifier*—A positive feedback work gate in which the amount of

2. H. T. Odum, *Environment, Power, and Society* (New York: Wiley-Interscience, 1971).

work applied at the top is necessary to increase by a constant factor the output of the system. It may represent such processes as reproduction, combustion (fire), and ecological succession.

(j) *Active Impedance*—A backforce that slows down energy flows. Examples include human clothing or insulation of any kind.

(k) *Adding Junction*—The intersection of two or more flows of energy of similar kinds that is additive.

The Structure of Energy Flow in Ecosystems

As noted earlier, the trophic, or feeding, structure of a biotic community can be described as a food pyramid, chain, or web (see figure 2.1). Any community will possess one or more hierarchies characterized by a series of producer-consumer, prey-predator relationships among its constituent populations. Figure 2.4 illustrates a food chain by use of energetic symbols. Within this structure energy is transferred from populations at one level in the chain or hierarchy to populations at the next level, and so forth.

There are three basic uses of energy in food chains. First, *production,* or the accumulation of biomass, at all levels. Second, *respiration* is the self-maintenance cost, in-

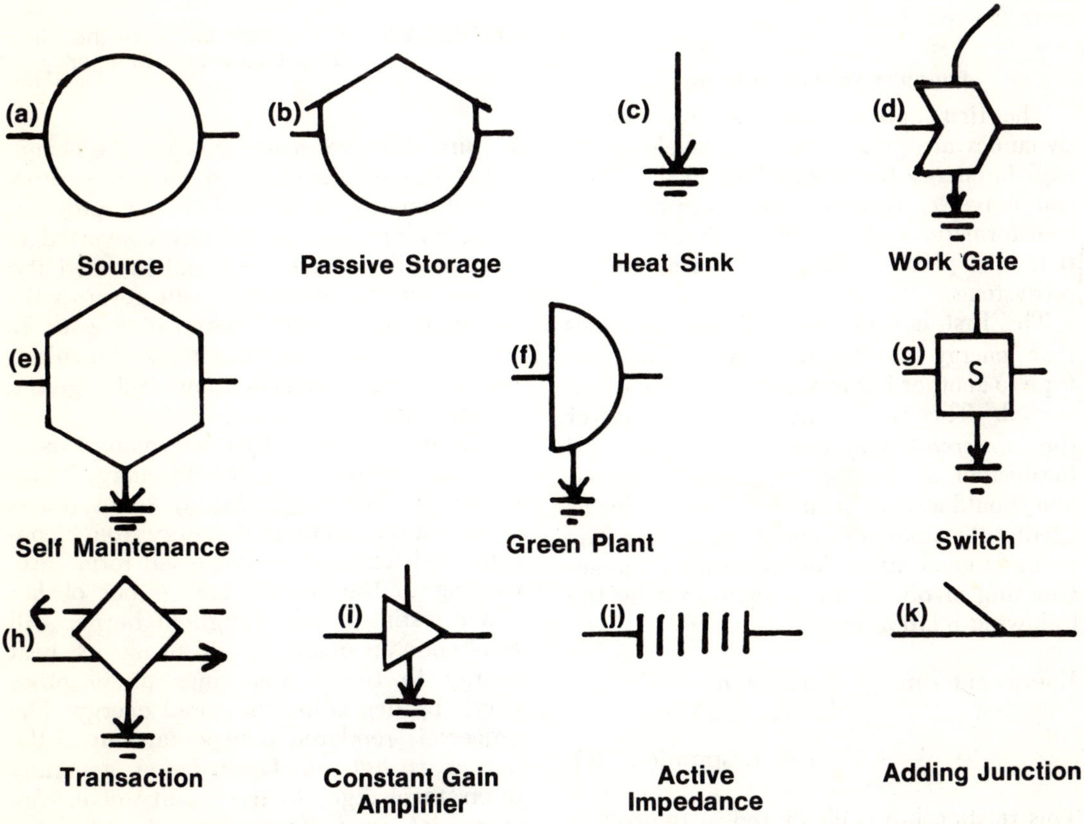

FIGURE 2.3. Modules of the energy language. (Adapted from Odum, H.T. 1971. *Environment, Power, and Society.* New York: Wiley-Interscience, p. 38.)

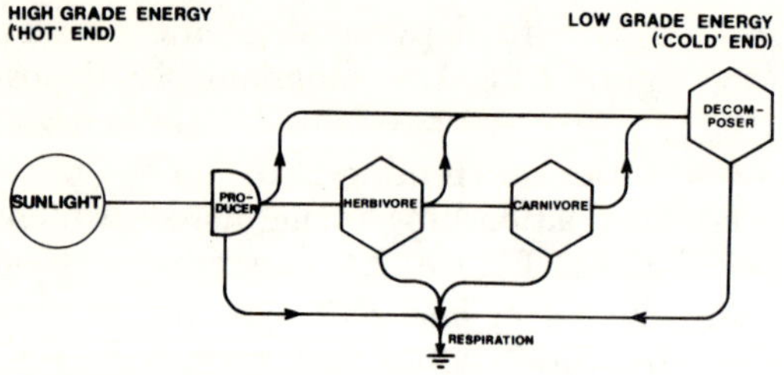

FIGURE 2.4. Primary and secondary food chains.

cluding growth, photosynthesis, or work, and includes the loss of heat as a by-product of energy conversion. Third, some energy is stored as dead organic matter or exported from the system.

Principles Relating to Energy

The first and second laws of thermodynamics are generalizations developed to explain energy transformations in the physical universe. However, they apply to all transformations of energy, including those that take place in biological forms and in ecosystems.

The first law of thermodynamics states that "energy may be transformed from one type to another but is never created nor destroyed." This is also known as the law of the *conservation of energy.* The practical implication of this law, or principle, is that one should always be able to account for or predict the amount of energy in a given system at a given time. For any system, the accounting involved can be expressed by the following relationship:

$$\text{Energy entering system (E)} = \text{Change in biomass } (\Delta M) + \text{Work (W)} + \text{Respiration (R)}$$

This relationship is illustrated in figure 2.5, and holds even if there are multiple stages or levels involving energy transformation in the system, as in a food chain. For example, sunlight is a form of kinetic energy transformed into potential energy by green plants through photosynthesis. This potential energy is in chemical form and may be converted further when herbivores use it.

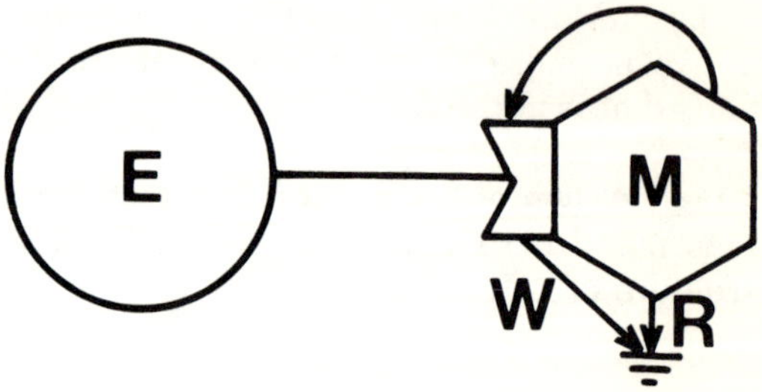

FIGURE 2.5. A representation of the "law of conservation of energy."

In turn, herbivores will convert the chemical energy of plants into other forms of potential energy, such as fats and proteins; some energy will be directly converted to heat in the course of metabolic activity; some energy is used in kinetic form as the herbivore moves about performing work. In such a chain of transformations, the rule is one of energy conversion and exchange but not of absolute loss.

The second law of thermodynamics states that "no process involving an energy transformation will spontaneously occur unless there is a degradation of energy from a concentrated form into a dispersed form." According to the second law, a pot of hot water resting on an unlighted burner will cool spontaneously by diffusing its heat energy to the surroundings or in other words by degrading its stored energy. The process is rendered nonspontaneous if the burner is lighted. Degradation generally refers to changes in the quality of energy from *high grade* (concentrated and utilizable) to *low grade* (diffuse, random, and unutilizable). This randomized, or low-grade state of energy—the ultimate fate of all energetic processes—is known as *entropy*.

Hence energy is also degraded as it passes through biological systems. However, this is not a spontaneous process, since biological systems require a more or less continuous input of energy to maintain them. Energy cannot be recycled, but the energy remaining after one transformation can be used for another. An approximate rule is that about 10 percent of the total energy available at the beginning of a biological transformation will be available for the next transformation in the food chain or pyramid. All such biological systems, then, have a "hot end" and a "cold end" (see figure 2.4).

Another derivation from the second law is referred to as the *stability principle*. This principle deals with the concept that any natural system through which energy flows tends to change until a stable adjustment, with *self-regulating mechanisms*, is achieved. "Stability" and "self-regulating mechanisms" refer to the manner in which the flow of energy, the cycling of nutrients, and other vital processes are *controlled* in biological systems at each level of the hierarchy. There are two kinds of control relevant to our understanding of the operation of biological systems. The first is the influence of *limiting* and *tolerance factors,* or *environmental controls.* The second, *internal feedback controls,* is under the influence of the activities and movements of components of the system.

CONTROL FACTORS

Environmental control of energy flow in ecological systems comes about through the influence of *limiting* and *tolerance factors* on the ability of components of the system to use available energy. This is because the growth and longevity of organisms are dependent on both the availability of energy and of other essential constituents. These include chemical elements and compounds and physical factors as well as the relative absence of hazardous substances. A variety of metalic and nonmetalic elements (including trace elements) and complex organic compounds can be limiting to organisms, including man. The availability of these substances to the system is not only a function of the geological and meteorological structure of the system but also of the interaction of these with the biotic components of the system. Earth and atmospheric characteristics also determine the prominence of such abiotic factors as water availability, temperature, and light. In general, limiting factors refer to constituents which are available in minimal quantities, whereas tolerance factors refer to overabundant constituents. The same constituent in an ecosystem may be a limiting factor for one population and a tolerance factor for another. Also, an element may be both a limiting and a tolerance factor for the same population depending upon the magnitude of that element. Vitamins, for example, can be limiting when intake is low but tolerance factors when intake is excessive.

The establishment of a particular population in a given area depends on the necessary substances in the required minimum quantity, the functioning of crucial physical factors at the required minimal levels, and the operation of these factors within the tolerance limits of the population. For example, in savanna grassland ecosystems, water availability is a limiting factor both to plant growth and to herbivores, and this limiting aspect has a strong seasonal basis. Moreover, because of the high evapotranspiration of water, salt deposits commonly form that in turn become tolerance factors to plant life. In effect, the operant adaptation of an organism can be viewed as its solution to past environmental problems relative to limiting and tolerance factors. A case in point is the high levels of resistance (or tolerance) that many savanna shrubs and grasses exhibit to soils with a high salt content. Figure 2.6 illustrates the action of limiting and tolerance factors on a food chain using energetic symbols.

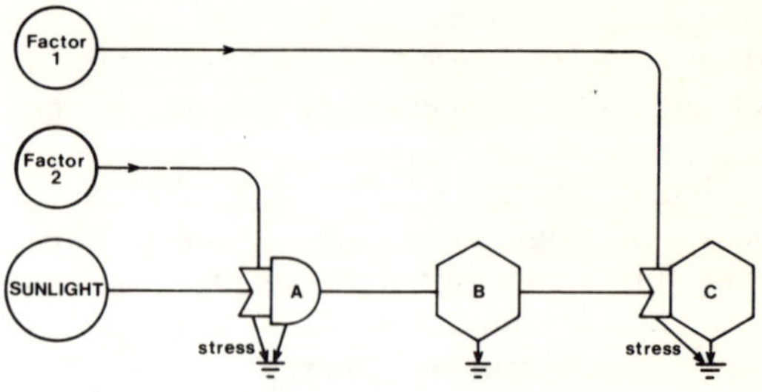

FIGURE 2.6. Environmental control of ecological systems involving the influence of limiting and tolerance factors. Component C has two environmental problems; its own (factor 1) and that of its community (factor 2).

Internal Controls and Homeostatic Processes

This general topic refers to the variety of mechanisms at the disposal of living systems, including organisms, populations, and communities, for dealing with environmental problems. These mechanisms involve feedback and control—the keys to the study of closed systems and homeostatic (equilibrium) processes (further discussed in chapter 4). The principle of control through feedback has great generality and is applied in man-made devices (machines), individual physiological processes, population control through reproduction and territoriality, and community-level processes such as predator-prey relationships.[3] Control-system logic has also been applied to human social phenomena and to man-environment interactions.[4] Feedback control is usually in operation in a given system when oscillations can be observed.

For feedback control to be demonstrated, three conditions must be met: (1) there must be some mechanism for producing the specified changes; (2) there must be a measuring device; and (3) both measurement and regulation must be of sufficient speed and strength to accomplish the task at hand.[5] Like a home thermostat, a specific characteristic (temperature) must be maintained within an appropriate range of values (66-70°F) despite changes in the environment (outdoor temperature). Failure of the temperature-control system in the home will lead to some discomfort, while failure of a feedback-control process in an ecosystem may jeopardize the existence of a population or community.

Slobodkin has proposed a hierarchy of homeostatic responses applicable to human cultural and biological systems.[6] He suggests that the primary responses to environmental problems (or stresses) are behavioral (least costly), where behavioral responses act as buffers against the necessity for physiological or biological changes, and where learning adds flexibility to behavior.[7] Similarly, physiological responses act as buffers against genetic change, with acclimatization adding flexibility to physio-

3. V. C. Wynne-Edwards, "Population Control in Animals," *Scientific American* 211, 2 (1964): 68-74.

4. R. F. Murphy, "Intergroup Hostility and Social Cohesion," *American Anthropologist* 59 (1957):1018-35; P. Collins and A. P. Vayda, "Functional Analysis and its Aims," *Australia and New Zealand Journal of Sociology* 5 (1969):153-6.

5. A. Tustin, "Feedback," *Scientific American* 187 (1952).

6. L. Slobodkin, "Toward a Predictive Theory of Evolution," in *Population Biology and Evolution*, ed. R. S. Lewontin (Syracuse: Syracuse University Press, 1968), pp. 187-205.

7. The cost of the response is the effect on the ability of the organism to respond to future problems.

logical responses. Genetic change, then, is the highest level response in this hierarchy but the slowest acting and most costly in terms of the loss of flexibility. Slobodkin also states that, "Even if the (environmental) perturbation is of an absolutely new kind in the animals' experience, . . . the initial response to it will be behavioral, since it will be coded as if it were one of the normal perturbations to which the animals can respond."[8] This is, of course, the ultimate test to which any species may be subjected; it describes the circumstances under which evolutionary change is most likely to occur.

From the standpoint of controlling the flow of energy and the cycling of materials in ecosystems, the primary homeostatic responses consist of the activities and movements of organisms and populations. In other words certain of these activities and movements are responses to perturbation in the availability of energy or other nutritional elements and can be ordered in a hierarchy of appropriate adaptive responses. In nonhuman animals these responses typically involve movement and population control. Humans, while employing movement- and population-control mechanisms, often have a variety of more powerful resource-management strategies at their disposal.

For Further Reading

Boughey, A. S. *Ecology of Populations.* New York: Macmillan Company, 1968. An introduction to ecology from a population standpoint, with special attention given to human ecology and evolution.

Energy and Power, Readings from Scientific American. San Francisco: W. H. Freeman and Company, 1971. A readable and broad collection of articles on energy, energetic principles, and energy in relation to man, including some case material discussed in this book.

Odum, H. T. *Environment, Power, and Society.* New York: Wiley-Interscience, 1971. An energetic approach to ecology with comprehensive discussion of application and use of energy-flow symbols.

Smith, R. L. *Ecology and Field Biology.* New York: Harper & Row, 1962. An excellent and thorough introduction to basic concepts and principles of general ecology, with special relevance to anthropologists because of its practical field orientation.

Watt, K. E. F. *Principles of Environmental Science.* New York: McGraw-Hill Book Company, 1973. This is a truly outstanding introduction to ecological principles. Considerable treatment is given to human populations.

Bibliography

Odum, E. P. 1959. *Fundamentals of Ecology.* Philadelphia: W. B. Saunders.

Patterns and Perspectives in Environmental Science, Report Prepared for the National Science Board, National Science Foundation. Washington, D.C.: National Science Foundation, 1972.

8. Slobodkin, "Toward a Predictive Theory," pp. 196-7.

3 Human Life-Support Systems

THE STRUCTURE OF HUMAN LIFE SUPPORT

For any human population, environmental problems usually arise from perturbations in the availability of energy and other nutritional substances. Human life-support systems try to provide the appropriate solutions. Those solutions to problems stemming from perturbations in the influence of physical tolerance factors and biotic hazards will be only partially considered within this framework.

Building on the fundamentals outlined in chapter 2, including the energy "language" and the principles relating to environmental and homeostatic control, we can now provide an outline of the structure and function of human life-support systems and integrate this with an outline of the structure and function of individual human physiology. The key to this is Slobodkin's proposition that "to a degree, behavior acts as a bulwark against the necessity for physiological change."[1] If so, then the biological condition of the members of a given population is in some sense a measure of the operation and success of the behavioral responses that are part of that population's life-support system. It is therefore necessary to describe the features of human life-support systems in some detail before discussing such biological matters as nutrition, reproduction, and aspects of disease.

The first task is to identify those culturally programmed human activities and movements that influence or buffer certain physiological responses. We have provisionally isolated two kinds of response structures. One is concerned with population control; the other involves resource-mangement strategies.[2] In their simplest form the latter involve changing the spatial dimensions of a population and its territorial food-acquisition activities. In more intensive versions, resource-management strategies involve environmental alteration, the integration of territories, and substitutes for human labor. We will propose and discuss a linking framework for these two types of response structures.

POPULATION CONTROL

In common with many animal species humans display a variety of demographic re-

1. L. Slobodkin, "Toward a Predictive Theory of Evolution," in *Population Biology and Evolution*, ed. R. C. Lewontin (Syracuse: Syracuse University Press, 1968), pp. 187-205.

2. A "resource-management strategy" is a set of practices which sustain the yield of a particular plant or animal source, or some other renewable natural resource. The phrase "sustain the yield" implies that resource-management strategies possess homeostatic properties.

sponses to environmental perturbation, responses which operate to control population through non-Malthusian devices. These devices are nevertheless closely correlated with food supply. Using Slobodkin's logic it is useful to view such behavioral responses as a buffer aginst Malthusian biological responses, or at least to posit a range of response levels with greater biological involvement from phase to phase. To promote a degree of generality we have constructed a demographic response hierarchy having five levels of response.

Level I: Social and Territorial Behavior

Included here are practices which have the effect of altering the size and structure of local human populations in relation to their energy and other dietary-element requirements as well as their reproductive potential. In mind are such mechanims as migration, and such institutions as witchcraft beliefs and social ranking (and its sometimes resultant stratification). Also included are: courtship practices; incest taboos and prescriptive or preferential marriage regulations; marriage rules (monogamy, polygany, etic.); institutionalized celibacy; social rules that artificially define sexual and social maturity by applying ritual, educational, and/or economic tests; and social mobility and class/caste endogamy in stratified societies. Such mechanisms are low in cost, and easily mobilized and reversed. Yet locally they can have the same magnitude of effects as birth (or in-migration) and death (or out-migration).

Level II: Reproductive Behavior

Included here are responses that modify the risk of exposure to conception. We have in mind such mechanisms as postpartum sex taboos and coital frequency. The latter can be affected by a range of factors including the organization of work, individual mobility, socialization, and the regular ingestion of toxic or narcotic substances. Also included at this level are coital and noncoital sexual techniques, genital and gonadal mutilation (subincision, infibulation, vasectomy, castration, hysterectomy), mechanical birth-control devices, and artificial hormonal regulation.

Level III: Gestation and Term

Included here are behavioral responses and biological responses that affect the probability of a successful pregnancy and live birth. Such behavioral responses as induced abortion and infanticide, and such physiological responses as adrenocortical stress, maternal nutrition stress, and birth defects may be included.

Level IV: Growth and Socialization

Included here as in Level III are behavioral responses and environmentally influenced biological responses, both affecting the degree and speed of achievement of physical and behavioral maturity. Biological responses—such as nutritionally controlled speed of growth and resultant stunting or tallness, early or late sexual maturity, training and fitness, and the acquisitions of partial immunities to infectious disease—as well as other types of adaptation can be incorporated. Behavioral responses pertaining to this level include those that affect the quality and intensity of child-care and such socialization practices as sleeping and feeding customs, the use of force, sexual indoctrination, traumatization, "programming," and the transmission of sex-role patterns. Because this level determines the physical and behavioral adaptability of future generations, it has the crucial role of restoring flexibility to lower level responses.

Level V: Longevity and Mortality

Length of life and mortality are the so-called Malthusian checks through which environmental, and to a lesser degree behavioral, links directly affect life expectancy and mortality rates through widespread starvation, epidemic disease, and high rates of homicide and suicide.

RESOURCE MANAGEMENT

Resource-management strategies are a distinctive feature of human behavioral responses to environmental problems. These function to alter the spatial dimensions of human life-support systems or control the rate of energy consumption of selected portions of ecosystems to support variable levels of human population. Traditional studies of human subsistence ecology tend to emphasize the energy consumption mechanism. Here and in later case studies, we argue that those mechanisms usually said to be the subsistence system of a human population give only a *partial* picture of the range of and the complex relationship between human activities and movements.

The historical reasons for the narrow focus of conventional studies of subsistence were explored in chapter 1. Julian Steward's definition of the "culture core" firmly fixed the preoccupation with technology and the application of labor to it. Subsequent studies that attempted to operationalize and quantify the study of "core" culture adhered closely to the original version. These studies involved collecting quantified data on the human labor input to narrowly defined agricultural, herding, and harvesting activities as well as data on the caloric outputs involved. Two erroneous conclusions resulted from this: (1) that the energetic efficiency of food production has improved in the course of cultural evolution; and (2) that the task of subsistence systems is to deal with the problem of the limited availability of nutritional energy.

Harris's work illustrates the first point, not because he is a strong proponent of the fallacies cited, but rather because of his advocacy of a materialist research strategy which forced him to base conclusions on poor or partial data or intuition.[3] The kind of labor input-food output data described permitted Harris to calculate what he called the *technoenvironmental efficiency* of food production; that is, the number of food calories produced per calorie of human labor input to food production. He was able to assemble data on five cases. These were intended to be broadly representative of different levels of social and technological complexity but were primarily determined by a paucity of useful data. With this admittedly limited and unrepresentative sample (see table 3.1) an attempt was made to generalize about long-term trends in food production systems.

> As the factor of technoenvironmental efficiency has improved, both the average size of the food-producing work force, and the total number of hours spent on food production has gone up . . . labor saving food producing technologies have been employed primarily to increase total production rather than to cut back on the work effort per food producer. Up until the introduction of industrial agriculture, this meant not only a continuous increase in work input per food producer but an even greater increase in the total number of food producers per square mile of arable land. Thus, in the long run, the increment in the techno-environmental efficiency has led to an increment in population density.[4]

As an examination of Harris's tabulated data will show, none of these conclusions relate directly to technoenvironmental efficiency. Rather the conclusions are based on a comparison of the size of the labor force and the hours of work in the various cases. The

3. M. Harris, *Culture, Man, and Nature* (New York: Thomas Y. Crowell Company, 1971).

4. Harris, *Culture*, pp. 216-17.

sample is insufficient to correlate statistically the increments in these dimensions to increments in technoenvironmental efficiency. For example, the conclusion that Harris draws relating technoenvironmental efficiency to population density is an intuitive one. What probably is happening is that as life-support systems become more complex and intensive, labor is diverted to non-food-producing but nevertheless life-supporting activities. This conclusion is at least suggested by the material presented in the case studies in chapter 5.

That something more is involved in making food and other necessary substances available to a population than growing and/or harvesting it and hence in calculating energetic efficiencies, has been suggested by some recent published work. Pimentel and associates attempted to calculate the "real" energy costs of modern U.S. agriculture, including the fossil-fuel subsidy and the human labor expanded to manufacture machinery, fertilizer, pesticides, and the like.[5] Steinhart and Steinhart similarly assembled data on the energy and labor cost of modern U.S. food transportation, processing, and packaging.[6] When all of these data are assembled, the notion of a reduction in human labor input and an increase in energetic efficiency of life support turns out to be invalid, at least as far as U.S. trends are concerned. In other words an investigation of how modern life-support systems work involves the quantification of farm-support labor in the activities specified, plus employment in food processing, wholesale, retail, and transportation firms, and in the manufacturing that supports them. As the Steinharts suggest, the "hamburger-slinging" fast-food service employee is as much a part of the U.S. life-support system as the farmer.

5. D. Pimentel et al., "Food Production and the Energy Crisis," *Science* 182 (1973):443-9.

6. C. Steinhart and J. Steinhart, *Energy: Sources, Use, and Role in Human Affairs* (North Scituate, Mass.: Duxbury Press, 1974), pp. 70-81, 330-40.

TABLE 3.1

THE SAMPLE AND DATA BASE FOR CALCULATING COMPARATIVE TECHNOENVIRONMENTAL (LABOR INPUT-FOOD OUTPUT) EFFICIENCIES

Population	(a) Production, kcal/yr.	(b) No. of Food Producers	(c) Man-hours per yr. per Food Producers	(d) Mean kcal Expended per hr.	(e) Techno-environmental Efficiency (a/bxcbd=e)
Bushmen	2.3×10^7	20	805	150	9.6
Genieri (Gambia)	4.6×10^8	334	820	150	11.2
Marings (New Guinea)	1.5×10^8	146	380	150	18.0
Marings (Pigs only)	1.8×10^7	146	400	150	2.1
China	3.8×10^9	418	1,129	150	53.5
U.S.A.	2.6×10^{14}	5 million	1,714	150	210.0

(After data from Harris, M. 1971. **Culture, Man, and Nature.** New York: Thomas Y. Crowell Company.)

Similar considerations are treated in the case studies in chapter 5. In particular, the case on the New Guinea Highlanders reflects an attempt to include all life-supporting activities for which data were available. We purposely erred in the direction of assuming that *most* work activities in all populations are life supporting. The framework that justifies this will be presented later in this chapter.

A second fallacy affecting the study of human life-support systems, cited earlier in this chapter, is the assumption that nutritional energy is a significant environmental problem (a limiting factor) for all human populations, with subsistence systems providing the solution to it. The *carrying capacity* concept and similar constructs (such as the "man-land balance"), as applied to the study of human ecological problems by geographers, anthropologists, and archaeologists, were based on a calculation of the caloric requirements of a given population and its comparison with a quantitative assessment of a given technology's ability in a particular environment to meet those caloric requirements.

A number of seemingly paradoxical conclusions were reached through this concept. For example, Carneiro found that certain human populations of lowland South America practicing shifting cultivation (slash and burn) were shifting their settlements even though an abundance of arable land remained closeby.[7] Although concluding that sociopolitical factors were at work, he more sensibly advanced the possibility that the abundance of game animals was a factor too. A similar example comes from Sahlins.[8] In reviewing primary data on tribal and peasant populations in which agricultural production appeared to be organized at the family level, he concluded that such systems were underproductive and not fully realizing their hypothetical potential. This was despite the fact that on a local population (suprafamilial) level caloric requirements were being met. Sahlins explained this by asserting that social structural factors were at work: that what he called the "domestic mode of production" consistently underuses its human (labor) and natural resources. Sahlins did not consider the possibility that expanding the production of calories would not promote the support of a larger population if some other substance or element, such as essential amino acids, was limiting to the population in question.

Another example is the changing view of pig husbandry practices in the New Guinea highlands. At one time it was considered to be a typical example of economic "irrationality," with New Guinea pig festivals cited in particular.[9] Then, on a speculative basis, Vayda and associates proposed a number of possible ecological functions for pig husbandry in New Guinea.[10] Receiving particular attention was the storage function of pigs, a solution to the problem of temporal variations in the availability of food. These authorities also noted the importance of domesticated pigs as a source of high-quality protein in an otherwise vegetarian diet, but the overall thrust of their argument was energetic. Rappaport's work with the Marings of New Guinea is significant to the

7. R. Carneiro, "Slash and Burn Cultivation Among the Kuikuru and its Implications for Cultural Development in the Amazon Basin," in *Man in Adaptation: The Cultural Present,* ed. Y. Cohen (Chicago: Aldine Publishing Company, 1968).

8. M. Sahlins, *Stone Age Economics* (Chicago: Aldine Publishing Company, 1972).

9. R. Bunzel, "The Economic Organization of Primitive Peoples" in *General Anthropology,* ed. F. Boaz (New York: D. C. Heath, 1938), pp. 327-408; R. Linton, *The Tree of Culture* (New York: Alfred A. Knopf, Inc., 1955), p. 98 ff.

10. A. P. Vayda, A. Leeds, and D. Smith, "The Place of Pigs in Melanesian Subsistence," in *Symposium: Patterns of Land Utilization and Other Papers,* ed. V. Garfield. Proceedings of the 1961 Annual Spring Meeting of the American Ethnological Society (Seattle: University of Washington Press, 1961), pp. 69-77.

STATE UNIVERSITY OF NEW YORK AT BINGHAMTON
Department of Philosophy

Please complete these forms. Please report all changes. Return to Secretary.

NAME: Robert Berra   HOME PHONE: 723-5147

ADDRESS: 16 Mendelssohn St.   OFFICE NO. 1111   PHONE: —

Bing., N.Y. 13905

List all classes, lectures & committees	Days & Hours	Room No.
Human Ecology	Th. – 5:30-8pm	SI 135
Semiotics	Tu. – 4-6:30 am	FA 241
German	MWF – 9:30-10:30	CW 214
Soc. Psych.	MTWTh. – 1:30-3:30	CW 215
Phenom. & Exist.	MW – 2:30-3:30	LH 9
	Disc. Sec. MW – 3:30-4:30	CW 306
~~Office Hours~~	~~TU 10-12am~~	

OFFICE HOURS: Tu. 10-12am, Th. 10-12 A.M.

present argument because he demonstrated the energetic inefficiency of pig husbandry.[11] This holds both from the standpoint of food-chain efficiency (the amount of food, otherwise edible by people, which is fed to pigs) and the human labor input-food output efficiency as compared with that for vegetable food production. This points to a general strategy of life-support systems—to allocate sometimes disproportionate amounts of human labor (and other work-generating resources) to those resource-management strategies which deal with the most pressing environmental problems (or limiting factors). This is done for reasons other than the achievement of a degree of energetic efficiency or the maximization of the output of nutritional energy. As a matter of fact, for a life-support system to work, the labor input-food output ratio need only be *less than* the worker-consumer ratio if all life-supporting activities are calculated. Within a given system the labor input-food output ratio is a variable inversely related to population pressure. Accordingly, it has little value for cross-cultural comparison but is valuable for measuring changes in the operation of a given life-support system in time.

Energy *is* limiting to some human populations. For preindustrial human populations this condition is directly related to latitude, with populations inhabiting Arctic and sub-arctic habitats, such as the Eskimos (described in one of the case studies) representing extreme cases of this. In addition, as suggested above, energy may be limiting in certain life-support systems under population pressure, or under the artificial or induced population pressure to be associated with rural-urban and colonial systems.

The general point is that although energy is in a sense basic, it is usually not limiting to most human populations. Instead, the utilization of available energy is controlled by the operation of other limiting and tolerance factors as well as by certain activities and movements of components of the system.

This necessary discussion provides a basis for talking about the kinds of resource-mangement strategies displayed by human populations, and how these may be arranged in hierarchies of response. Four broad types of resource-management strategies designate ways in which members of human populations may act to control the flow of energy and the availability of other substances for the life support of their populations. These types are ordered in a hierarchy that can also represent a developmental sequence for life-support systems in time. The four types of resource-mangement strategies are outlined below.

Level I: Movement Strategies

These strategies involve the internal movement of individuals and subgroups of a given population within its territory. In the usage adopted here it is not synonymous with migration. Migration involves permanent or long-term (reversible) changes in population membership. Movement involves locating individuals within a particular habitat and a given dispersal area in relation to food resources and/or in connection with reproduction or some other essential purpose. Movement patterns tend to be cyclical, related to oscillations in natural or man-influenced environmental parameters. The highest level of response in a structure of movement strategies is warfare or some other form of group violence. Warfare is movement because it involves either an attempt to extend or defend the amount of space in which other strategies are performed. Hence warfare or group conflict may involve the intensification of movement strategies to restore the flexibility of

11. R. A. Rappaport, *Pigs for the Ancestors* (New Haven, Conn.: Yale University Press, 1968).

other strategies. Typically, those human populations which anthropologists designate "hunters-gatherers" primarily use movement strategies to deal with environmental problems. Local populations of hunter-gatherers are not sedentary. They respond to perturbations in the availability of limiting resources and the operation of tolerance factors by moving. Populations commonly invade pristine or recovered habitats, they periodically nucleate or disperse, and individuals participate in more idiosyncratic patterns of movement. Among the human populations described in the case studies, the Arctic Eskimo depended exclusively on movement strategies for resource management until recent times.

Level II: Environmental Alteration Strategies

The spatial organization of habitats is such that most resources are scattered. As a consequence, populations adhering primarily to movement strategies expend most of their labor in finding these resources. This is what movement is all about. Environmental alteration strategies deal directly with the problem of scattered resources by directing increasing amounts of human energy expenditure into reducing the scatter. In other words biotic resources are made more available by artificially increasing their order and predictability (reducing entropy); this requires higher inputs of energy usually in the form of human labor. The effect is to create artificially favorable habitats for particular plant and animal species. Since domesticates and cultigens are almost invariably removed from their original habitats, the domesticated species are almost completely dependent on man. The relationship between man and domesticate or cultigen is one of "mutualism"—both benefit. The range of techniques which may be employed in the agricultural and pastoral variants of this kind of strategy are extremely diverse. Some are discussed in detail in appropriate case studies.

Typically, those human populations which anthropologists designate "tribal" and "agricultural" or "pastoral" conform to this general pattern of response, practicing environmental alteration strategies along with movement strategies. Such groups respond to population pressure, first by expanding or extending the spatial application of existing strategies to the limit (a form of movement), and then by intensifying their environmental alteration strategies. They do this by adding strategies, that is, those which usually entail increasing the amount of labor or other sources of work required in their operation, with resulting increments in the amount of energy and other substances which may be extracted from a given area of land.

In the case studies, all of the groups discussed, with the exception of the Eskimos, employ environmental alteration strategies. The East African Pastoralists, for example, practice a mixture of movement and environmental alteration strategies. The former involves the movement of people and herds over the range as well as individual and subgroup mobility between larger groups. The latter involves mangement of reproduction and exploitation of the herds and resorting to agricultural strategies when abiotic factors permit.

Level III: Areal Integration Strategies

The two lower-level kinds of strategies discussed (Movement and Environmental Alteration) involve a form of local sufficiency or economic encapsulation with respect to consumable goods, especially food. Under an areally integrating strategy, locally based human populations lose self-sufficiency (and ultimately social and political self-determination) and are dependent on extensive networks of economic exchange for some of their limiting resources. These networks integrate wide and diverse areas for life support and promote the maintenance of higher population levels. Another way of putting this is that some local populations are able to increase their numbers *as a consequence of* their participation in the network. This is because local limiting factors are transcended and population is influenced by the limiting factors pertaining to the larger integrated area. Populations may become more sedentary with the adoption and intensification of areally integrating strategies. To an extent movement strategies become submerged as labor and energy is expended to move essential substances to the people rather than moving the people to the sources of these substances. This also implies that a given local population imbedded in such a network will abandon certain strategies and become more specialized.

The sociopolitical dimensions of areally integrating strategies are particularly prominent. The intensification of such strategies involves the elaboration of political roles, ranging at the lowest level from the "big men" of the New Guinea highlands who have a principal role in controlling the movement of pigs in a small local portion of wide ranging networks to the kings and presidents who stand at the apex of the control structure in state-organized societies. A concomitant of this intensification is the increasing prominence of force, or its threat, in connection with control and maintenance of the work gates involved in the life-support system. Another feature is the creation of central places; ritual and festival centers, markets, and cities.

The case studies illustrate a number of these points. In the Altiplano Indian case study one should note the relationships between the Indian, the *hacienda* owner, and the larger Peruvian state. To gain access to land, and hence life support, the Indians must work for the hacienda owner and surrender a significant proportion of their production. The relationship is supported by the state which typically would resist with force an attempt to change its character. Indians in the highlands of South and Middle America have occasionally resorted to the use of force themselves in an attempt to "capture" the work gate involved. It should be noted that it was not feasible to represent all of these functions in the energy-flow model accompanying the case study. The life-support benefits of areally integrating strategies are also illustrated in the case studies. For the Altiplano Indians, the production of wool and its exchange for food from lowland areas appears to be an efficient strategy, at least from the standpoint of nutritional returns. Similarly, the participation of New Guinea highland populations in traditional and modern exchange networks compensates directly for the influence of the limited availability of fats and protein.

Level IV: Industrial Strategies

Industrial strategies involve the substitution of work from nonhuman animals and nonliving energy sources *to operate the other kinds of strategies.* The simplest form of this involves the use of animals for traction and as beasts of burden. Other lower-level versions of the strategy involve the small-scale harnessing of the kinetic energy of wind and water to transport nutrients (as in irrigation), or goods (as on canals or in sailing ships), and to drive simple machinery (windmills, waterwheels, and the like to drive pumps and other basic machines). The most intensive versions of industrial strategies involve the use of fossil fuels and nuclear fission for energy to perform work for personal mobility, agriculture, and the transport of goods.

A general point requiring emphasis is that all human populations do not have resource-mangement strategies representative of all the types or levels. In addition, the strategies constituting the life-support system of any population may themselves be arranged in a response hierarchy. Practically, it seems feasible to do this only for a fairly narrow set of strategies. For example, this has been done for Miyanmin (New Guinea) movement strategies.[12] Vayda has done this for the war process among the Maori (New Zealand) and the Marings (New Guinea).[13]

Having provided an outline of the operation of two apparently distinct response structures, we can now briefly propose a method for integrating them. A combination of Boserup's hypothesis and Slobodkin's logic suggests that the population-control response structure and the resource-mangement response structure bear a hierarchical relationship to each other that is charcterized by positive feedback.[14] This is represented in figure 3.1. Thus responses which have their homeostatic effect by managing resources, serve within limits to buffer demographic responses. When those limits are reached, demographic responses restore flexibility to resource-mangement responses by acting within limits to reduce population pressure. This is accomplished by providing the conditions that favor the adoption of behavioral innovations, including those involving more intensive use of the environment. Specifically, these conditions involve higher levels of mortality. The adoption of a resource-management innovation in turn, may restore flexibility to the hierarchy of demographic responses, with that portion of the system reverting to lower levels of response.

The relationship between ecosystem structure and function and the controlling activities and movements of populations of plants, animals, and humans is given in figure 3.2. The controlling activities and movements constitute the life-support system for a given population. The half-brackets enclose the kinds of life-support strategies employed by human populations in contrast to those employed by plants and animals. Arrows indicate the points in the ecosystem structure at which the controlling effects of the various strategies are imposed.

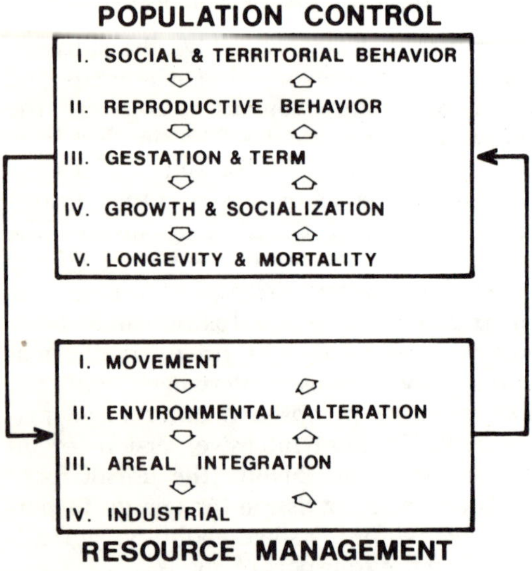

FIGURE 3.1. A general model of the hierarchy of responses to life-support stress.

12. G. E. B. Morren, *Settlement Strategies and Hunting in a New Guinea Society*, Ph.D. diss. (New York: Columbia University, 1974).

13. A. P. Vayda, "Maoris and Muskets in New Zealand: Disruption of a War System," *Political Science Quarterly* 85 (1970):560-84; A. P. Vayda, "Phases in the Process of War and Peace Among the Marings of New Guinea," *Oceania* 42 (1971): 1-24; A. P. Vayda, "Warfare in Ecological Perspective," *Annual Review of Ecology and Systematics* 5 (1974): 21-31.

14. E. Boserup, *The Conditions of Agricultural Growth* (Chicago: Aldine Publishing Company, 1965); L. Slobodkin, "Toward a Predictive Theory of Evolution," in *Population Biology and Evolution,* ed. R. C. Lewontin (Syracuse: Syracuse University Press, 1968), pp. 187-205.

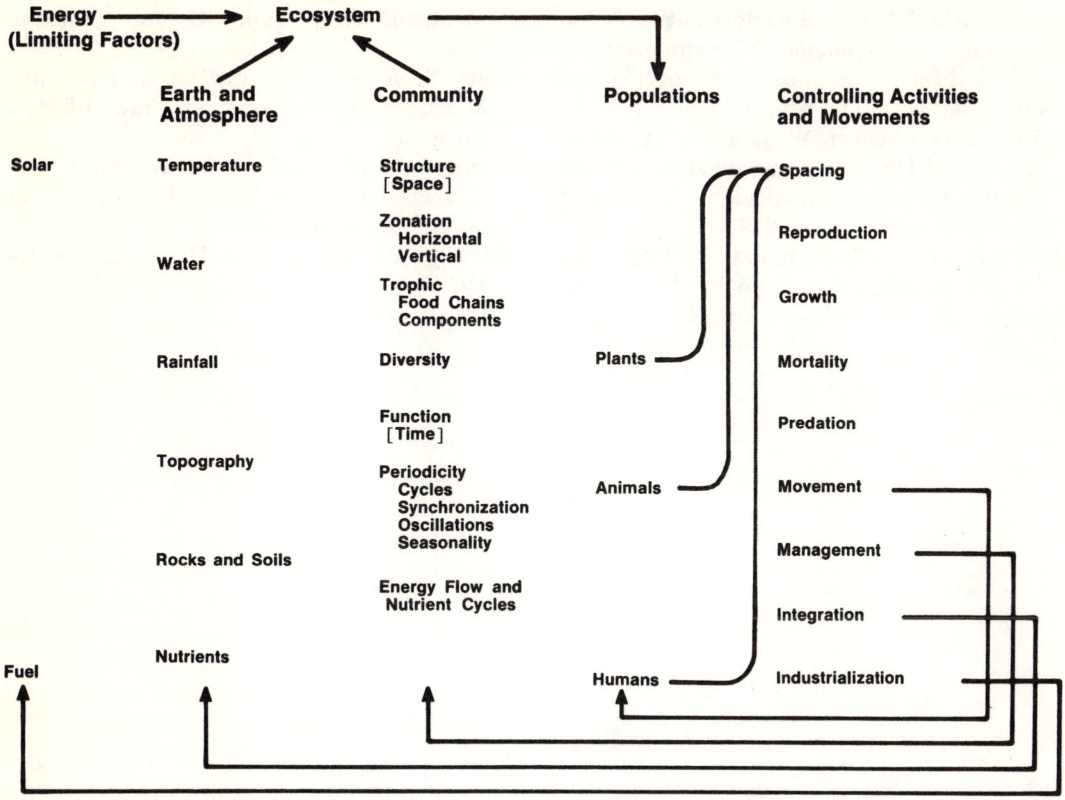

FIGURE 3.2. A model of ecosystem characteristics and controlling activities and movements.

For Further Reading

Boserup, E. *The Conditions of Agricultural Growth: The Economics of Agrarian Change Under Population Pressure*. Chicago: Aldine Publishing Company, 1965. A seminal work presenting an anti-Malthusian theory of the relationships between changes in population and resource utilization.

Malthus, T. "A Summary View of the Principle of Population," in *Three Essays in Population*. New York: Mentor Books, 1960. A synoptic view of Malthus's classic work.

Spooner, B., ed. *Population Growth: Anthropological Implications*. Cambridge, Mass.: MIT Press, 1972. A collection of papers exploring the implications of Boserup's theory, with particular attention to the role of population pressure triggering changes in life-support strategies.

Bibliography

Calhoun, J. B. 1952. "The Social Aspects of Population Dynamics." *Journal of Mammalogy* 33:139-59.

Christian, J. J. and Davis, D. E. 1964. "Endocrines, Behavior, and Population." *Science* 146:1550-60.

Davis, K. and Blake, J. 1956. Social Structure and Fertility: An Analytical Framework." *Economic Development and Cultural Change* 4:211-35.

Katz, S. 1972. "Biological Factors in Population Control." In *Population Growth: Anthropological Implications,* ed. B. Spooner. Cambridge, Mass.: MIT Press.

Schrire, C. and Steiger, W. L. 1974. "A Matter of Life and Death: An Investigation into the Practice of Female Infanticide in the Arctic." *Man* (new series) 9:161-84.

Stott, D. H. 1962. "Cultural and Natural Checks on Population Growth." In *Culture and the Evolution of Man,* ed. M. F. Ashley-Montagu. New York: Oxford University Press.

Watt, K. E. F. 1968. *Ecology and Resource Management.* New York: McGraw-Hill Book Company.

Wynne-Edwards, V. C. 1962. *Animal Dispersion in Relation to Social Behavior.* Edinburgh: Oliver and Boyd.

Zelinsky, W. 1971. "The Hypothesis of the Mobility Transition." *Geographical Review* 61:219-49.

4 Human Variability

INTRODUCTION: STRESSES AND RESPONSES IN HUMAN SYSTEMS

Biological systems, whether viewed as individual organisms, populations, or communities, must maintain some degree of *homeostasis*. Homeostasis means a steady state, or more correctly, a dynamic equilibrium held within certain limits. As noted in chapter 2, such a dynamic equilibrium is sustained by various *feedback mechanisms* that control the steady state and maintain it at a given level. The temperature of the human body is an example of homeostasis at the individual organism level; deep body temperature cannot go beyond certain limits if the individual is to survive. At the population and biotic community levels, numbers of individuals of a given species must be maintained within certain limits to avoid disrupting the balance of energy and material flows through the community and the ecosystem. Any force or action that produces a deviation from homeostasis is called a *stress* (or *stressor*), and the actual deviation from homeostasis is identified as *strain*.[1] Restoration of homeostasis and the consequent reduction of strain is an *adaptation* or adjustment to the original conditions of environmental stress. Adaptations to environmental stresses can be short-term, prolonged, or evolutionary in scope, and such adjustments are a basic property of all living systems, including human biological, behavioral, and life-support systems.

The following discussion focuses on a variety of stresses that arise from the abiotic and biotic components of ecosystems as well as from the human-population component. These include climatic, nutritional, disease, and demographic forms of stress, and the corresponding patterns of adaptive responses that have both biological and cultural bases.

CLIMATIC STRESS

Temperature, solar radiation, low oxygen pressure, and seasonal and daily variations constitute the climatic forms of stress pertinent to our discussion.

Temperature

Like other mammals, humans are *homeotherms* and are able to regulate their internal body temperature at levels around 37°C (98.6°F). The maintenance of this body temperature requires a complex pattern of controls of heat production in the body, transfer of heat within the body, and dis-

1. G. E. Folk, Jr., *Textbook of Environmental Physiology*, 2d ed. (Philadelphia: Lea and Febiger, 1974), p. 402.

sipation (loss) of heat from the body to the external environment. In a resting, unclothed individual any environmental temperature very much above or below 28°C (82°F) will require the activation of controls to modify heat production, transfer, and dissipation. Since humans are at rest for only a part of each day, and since environmental temperatures throughout the biosphere are seldom at the "ideal" level of 28°C, the human body must almost continually be regulating the body temperature to levels compatible with life and normal function. Temperature stress from the external environment begins to influence the human system when environmental temperatures become cold or hot, that is, deviate substantially from 28°C.

Under cold conditions (polar zones, mountains, temperate zones during winter, and other areas at night), the primary need is to prevent heat from being lost from the body surface. This is best accomplished by insulation in the form of clothing and shelter, although a variety of human biological factors can also serve to reduce heat loss (or in fact increase heat production). Under hot conditions (in the wet and dry tropics and temperate zones during the summer), the primary need is the dissipation of excess internal heat through the surface of the body. Here a distinction must be made between hot-wet and hot-dry climates. In hot-dry climates evaporative cooling through the sweating mechanism can effectively cool the body under all but the most extreme conditions.[2] However, one problem in a hot-dry environment is that water resources are frequently scarce, and thus may limit the ability of an individual to replenish water reserves in the body. In hot-wet climates with high humidity the sweating mechanism is less effective, and other modes of heat loss may play important roles. Contact with solid objects at a different temperature (*conduction*), the movement of air across the skin (*convection*), and the transfer of heat by infrared or heat radiation (*radiation*) can all influence the loss (or gain) of body heat.

Metabolic heat production is an important by-product of the biological processes that take place in the body; it can be increased or decreased in humans under a variety of conditions. Even the minimal level of metabolism (*basal metabolism*) shows human-population variation apparently linked to climatic temperatures. Populations living in hot climates tend to have relatively low basal metabolism, while those in cold climates tend to have slightly elevated basal metabolism.[3] Roberts has shown this relationship between mean annual temperature and basal metabolic rate in a group of 160 human populations (see figure 4.1).[4] The increase in basal metabolism among residents of cold zones is best exemplified by the Arctic Eskimo, the Alacaluf of Tierra del Fuego, and the Andean Indian. There are nevertheless absolute lower limits to basal metabolism, and dwellers in hot climates can only approach this lower limit of heat production by behavioral patterns of inactivity. In the cold, heat production can be raised dramatically by muscular activity (exercise) and by shivering, although there is a cost in the form of increased nutritional (caloric) requirements.

2. The evaporation of 1 gram of water from the skin removes about 0.58 kcal of heat from the surface of the body.

3. S. Robinson and D. L. Wiegman, "Temperature and Humidity, Part B: Heat and Humidity," in *Environmental Physiology*, ed. N. B. Slonim (St. Louis: C. V. Mosby Company, 1974), pp. 84-118; L. D. Carlson and A.C.L. Hsieh, "Temperature and Humidity, Part A: Cold," ibid., pp. 61-84.

4. D. F. Roberts, "Climate and Human Variability" (Reading, Mass.: Addison-Wesley Publishing Company Module in Anthropology no. 34, 1973).

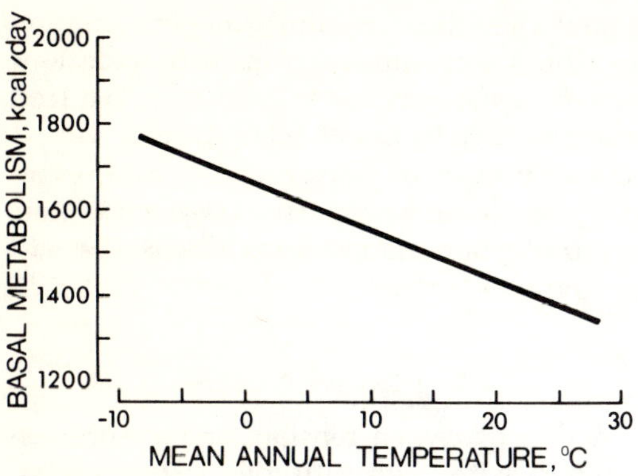

FIGURE 4.1. The relationship between mean annual environmental temperature and average basal metabolic rate in 160 populations. (Adapted from Roberts, D.F. 1973. "Climate and Human Variability." Reading, Mass.: Addison-Wesley Module in Anthropology no. 34, p. 6.)

Individual body size and proportions in human populations, although structured by processes of growth and natural selection, appear to be related to climate and environmental temperature variations. Since regulation of body temperature depends to a great extent upon preventing or aiding heat loss from the body surface, there are optimal models of body size and morphology (form) to prevent heat loss in the cold and aid heat loss in the heat. This relates to the surface area:volume (or weight) ratio of the human body. Small and linear (thin) individuals have a relatively high surface area: small, thin persons will be able to lose heat more readily than large, rotund individuals. Large and rotund individuals, conversely, will have a relatively low surface area:weight ratio and will be best able to conserve body heat by minimizing loss at the skin surface. Hence the short and linear individual will be best adapted to the tropics, while the large and rotund individual will be better adapted to colder climatic zones. Roberts again has statistically demonstrated these relationships with data from large numbers of populations throughout the world.[5] There is also some experimental evidence from laboratory studies of lean and obese women that confirm these relationships.[6]

Solar Radiation

Of all the solar radiation that strikes the surface of the earth, about half is *infrared;* the remainder is *ultraviolet* and *visible* radiation.[7] The influence of these three types of solar radiation on living forms, including humans, is manifold. Infrared and most visible radiation have a heating effect that can modify environmental temperatures and place an additional heat load on humans working in direct sunlight. Ultraviolet and some visible radiation have direct biological effects on plants and animals: (1) visible light provides energy for photosynthesis in green plants, and regulates biological rhythms through photoperiod effects; (2) ultraviolet radiation is involved in the production of vitamin D in human skin, has a germicidal effect on microorganisms, and can damage poorly protected living tissue.

Cloud cover, other atmospheric elements, and vegetational cover profoundly effect the amount of solar radiation that impinges on man at ground level. Desert ecosystems in the tropics afford litle cloud cover or shelter; both direct and *ground-reflected* (from light-colored soils) heat and ultraviolet ra-

5. Ibid.

6. J. Kollias et al., "Metabolic and Thermal Responses of Women During Cooling in Water," *Journal of Applied Physiology* 36 (1974):577-80.

7. F. Daniels, Jr., "Radiant Energy, Part A: Solar Radiation," in *Environmental Physiology,* ed. N. B. Slonim (St. Louis: C. V. Mosby Company, 1974), pp. 276-87.

diation are intense. Radiant levels are relatively high in mountain zones because the rarified atmosphere absorbs less direct radiation. Even snow fields, which have high reflecting properties, are characterized by moderate levels of radiation. Forests, especially tropical rain forests, give greatest protection from solar radiation. Dense canopies filter out a large percentage of direct sunlight, and the little sunlight that penetrates the forest growth is partly absorbed by the ground-level vegetation.

It is clear from what we know of the charcteristics of human skin that the dark *melanin* pigment present in varying degrees in all human populations serves a protective function against the damaging effects of solar radiation. Lightly pigmented skin with the ability to tan is common in populations from temperate zones, while tropical populations tend toward darker pigmentation. There are, however, elements of temperature and of ultraviolet radiation adaptation in the worldwide distribution of skin color in man. For example, Baker demonstrated that heavy skin pigmentation, as opposed to moderate pigmentation, increases the heat stress imposed on men working under desert conditions.[8] A light layer of clothing tends to reduce the heat load from solar radiation and probably explains the practice of wearing full robes in the arid zones of the Near East. Thus the advantage of heavy pigmentation to screen out ultraviolet radiation must be balanced against the disadvantages of an increased heat load in the skin pigmentation characteristics of any given population. The short-term effect of intense ultraviolet radiation in light-skinned individuals is sunburn, which can be debilitating if subsistence practices require prolonged outdoor work. Long-term repeated exposure to ultraviolet radiation leads to several forms of skin cancer, some of which are more serious than others. Finally, although highly speculative (and not directly related to solar radiation), it has been suggested that the melanin skin pigmentation distribution in human tropical populations may in some way be related to a reduced susceptibility to insect bites. Insects are, of course, vectors of numerous diseases, especially in humid tropical ecosystems, and any protection against such diseases would be highly adaptive.

Low Oxygen Pressure

Reduced oxygen tension, or pressure, associated with high altitudes is one of the most severe forms of climatic stress experienced by man. Other forms of climate stress can be buffered by cultural means, but there are no practical cultural means of reducing the stress of low oxygen pressure for the millions of people who reside in high mountain zones. Today there are probably twenty-five million people who live at elevations greater than 2,000 meters (6600 feet), and twelve million people who live above 3,000 meters (9,800 feet) in the Himalayas, Andes, Ethiopian highlands, and Caucasus mountains.

The primary effect of lowered oxygen pressure as felt by humans and other animals is *hypoxia*. Hypoxia is a state where insufficient amounts of atmospheric oxygen fail to satisfy the needs of body tissues: the effects are felt by nearly all systems in the body. A visitor at 3,500 meters (11,500 feet) will commonly experience headaches, loss of appetite and nausea, shortness of breath, mental impairment, and visual problems. These symptoms are known collectively as *soroche*, or mountain sickness, and should subside with acclimatization after about a week.[9]

8. P. T. Baker, "Racial Differences in Heat Tolerance," *American Journal of Physical Anthropology* 16 (1958):287-305.

9. C. M. Monge and C. C. Monge, *High-Altitude Diseases: Mechanism and Management* (Springfield, Ill.: C. C. Thomas, 1966).

The causes of hypoxia can be illustrated by an inspection of figure 4.2, a graph of the oxygen (O_2) dissociation curve of hemoglobin (Hb). Oxygen is inhaled into the lungs where it passes into the bloodstream by being bound to hemoglobin, the primary oxygen-carrying substance in red blood cells. At sea level hemoglobin becomes about 97 percent saturated with oxygen; each 100 milliliters (ml) of blood can carry

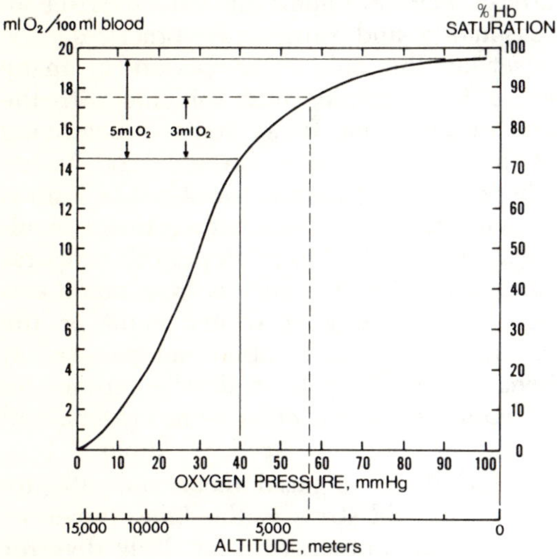

FIGURE 4.2. The oxygen dissociation curve of hemoglobin showing the variation in amounts of oxygen that can be carried by the blood at different oxygen pressures.

about 19.5 ml of oxygen. This occurs at an oxygen pressure in the blood of about 100 mm Hg (mercury). When the blood gives up its oxygen to the tissues, the oxygen pressure drops to about 40 mm Hg, and about 5 ml of oxygen for each 100 ml of blood has been delivered. The dashed line indicates what happens in an individual who is newly resident at 4,000 meters (13,100 feet) above sea level. In this case an oxygen pressure in the blood of only 57 mm Hg enables the blood to be less than 90 percent saturated with 17.5 ml of oxygen carried by each 100 ml of blood. The pressure drop to 40 mm Hg allows only 3 ml of oxygen to be delivered to the tissues, a value which is 60 percent of that delivered at sea level.

The short-term accommodations to hypoxia made in the visitor and the long-term accommodations manifest in the altitude native are complex and beyond the scope of this brief discussion.[10] However, they involve respiratory changes, increases in hemoglobin content, alterations in the blood circulation system, and modifications of the blood and tissue biochemistry. Populations with a long history of residence at high altitude are not immune to the effects of hypoxia under extreme conditions but are nevertheless remarkably tolerant to this form of stress. This is especially true during strenuous exercise which increases hypoxic stress.[11]

Seasonal and Daily Variations

Nearly all organisms display biological and behavioral rhythms with seasonal (*circannual*) and daily (*circadian*) periods that are synchronized with corresponding fluctuations in the environment. Photoperiod (light and darkness cycles), environmental temperature, precipitation, and a variety of other cyclic phenomena are reflected in the

10. For a detailed discussion of the mechanisms and patterns of adaptation to high-altitude stress, see, Wm. A. Stini, *Ecology and Human Adaptation* (Dubuque, Ia.: Wm. C. Brown Company Publishers, 1975), pp. 53-64.

11. P. T. Baker, "Human Adaptation to High Altitude," *Science* 142 (1969):1149-56; A. Hurtado, "Animals at High Altitude: Resident Man," in *Handbook of Physiology, Section 4: Adaptation to the Environment*, ed. D. B. Dill, E. F. Adolph, and C. G. Wilber (Washington, D. C.: American Physiological Society, 1964), pp. 843-60.

rhythmic patterns of organisms, including humans. Photoperiod variations can be best viewed by contrasting high-altitude zones (polar) with low-latitude zones (equatorial). At the equator there are no seasonal variations in daylight and night; throughout the year the 24-hour day is equally divided between light and darkness (L:D ratio is 12:12). At the poles of the earth the opposite case holds where seasonal variation is great with continuous daylight for six months (L:D ratio is 24:0), continuous darkness for six months (L:D ratio is 0:24), and virtually no daily variation in either of the two seasons. Middle latitudes (temperate zones) are intermediate, with seasonal photoperiod cycles consisting of longer light periods during the summer months and longer dark periods during the winter months.

Environmental temperature cycles tend to follow solar photoperiod cycles. In the tropics daily temperature fluctuations tend to be greater than seasonal ones; in the Arctic seasonal changes are of greater importance. Temperate zones thus tend to show both seasonal and daily cycles. There are innumerable minor exceptions to these generalizations based on special climatic conditions and ecosystem characteristics, but the basic pattern holds for much of the earth. One important exception is the fact that inland temperate zones show greater seasonal extremes than coastal zones at comparable latitudes. This is based on the mediating effect of large bodies of water on climate.

Seasonal patterns of precipitation, common to the tropics, have different effects in relatively wet and relatively dry ecosystems. In tropical rain forests where precipitation is high throughout the year and especially high for a few months, flooding of large areas will force human populations into a migratory pattern that coincides with seasonal rainfall. Arid tropical zones on the other hand may have amounts of rainfall that would be adequate for substantial vegetational growth if it were evenly distributed throughout the year. However, since rainfall is moderate to heavy for only a few months, the ecosystem is limited by the longer period of dryness. Such rainfall patterns in African savanna also require movements of human populations in quest of water.

The stresses on human populations from periodic fluctuations of climatic variables are especially important with respect to seasonality and subsistence practices. The onset of the cultivation period in many areas, for example, must coincide with the end of nighttime frosts and the beginning of the rainy season, as in the central Andes. There are also seasonal variations in human activity and nutritional intake that can ultimately be linked to photoperiod, temperature, and rainfall. There is even some evidence that the long winter night of the Arctic contributes to abnormal patterns of behavior, such as time disorientation and pronounced mental changes in visitors, and hysterical states in Arctic Eskimos.[12] Seasonal rhythms in basal metabolic rate are related to cold stress in the Ama, Japanese, and Korean women who regularly dive for seaweeds and shellfish in coastal waters.[13] During the summer months sea water reaches about 28°C (82°F) and basal metabolism (heat production) is at its lowest level; in the winter water temperature drops to about 10°C (50°F) and basal metabolism increases to nearly 40 percent over summer values. This appears to be a good example of a seasonally based adaptation to extreme cold.

12. H. W. Simpson and J. G. Bohlen, "Latitude and the Human Circadian System," in *Biological Aspects of Circadian Rhythms,* ed. J. N. Mills (London: Plenum, 1973), pp. 85-120.

13. S. K. Hong, "Comparison of Diving and Nondiving Women of Korea," *Federation Proceedings* 22 (1963):831-33.

Circadian rhythms of body temperature have been studied extensively. In most individuals oral or rectal temperature peaks in the late afternoon or early evening and is at its minimum point in the early morning. With this in mind, it should be clear that heat stress or cold stress affect the human body differently depending on the time of day or night. In one study, for example, it was found that men exposed to hot temperatures in the laboratory tended to be most resistant to heat stress in the early hours of the morning.[14] Although there is a distinct relationship between circadian body temperature rhythm in humans and environmental rhythms, the adaptive significance, if any, of this relationship is still not clear.

NUTRITIONAL STRESS

Humans require a variety of substances in the diet to survive, function, and maintain optimal health. These include: *water; energy and nutrients from proteins, fats, and carbohydrates; vitamins; and minerals.* The content and amounts of these intakes are dependent on such factors as age, sex, body size, body composition, physical activity patterns, state of health, prior nutritional state, and environmental temperature.[15]

Water content of the human body varies according to age and fat content. In newborn infants, water consitutes about 75 percent of body weight; in adults, about 50-60 percent. Water as a percentage of body weight declines with increasing fat (adipose) tissue; hence women, on the average fatter than men, have less body water. The daily rate of water turnover, equivalent to daily water requirements, is about 6 percent. This amount of course increases substantially under hot-wet or hot-dry environmental conditions, during strenuous exercise, and when an individual has a fever. In these cases heat must be dissipated by sweating and evaporative cooling at the expense of increased water requirements. Salt requirements (as sodium chloride) are increased as a result of heavy sweating, since salt is excreted in large amounts in sweat.

Basic energy requirements from proteins, fats, and carbohydrates vary enormously from person to person and from population to population. Energy intake from food must, of course, be in balance with energy expenditures or the human body will either gain or lose body weight. The energy requirements to maintain such a balance are illustrated in figure 4.3. Basal metabolism

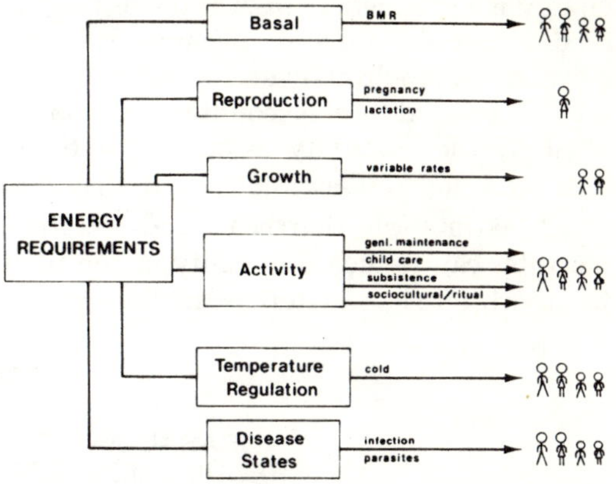

FIGURE 4.3. Basic energy requirements for humans.

uses about half the energy required, although this value increases in younger children and residents of cold climates where basal metabolic rates are higher. Energy requirements and especially protein re-

14. M. A. Little and J. A. Rummel, "Circadian Variations in Thermal and Metabolic Responses to Heat Exposure," *Journal of Applied Physiology* 31 (1971):556-61.

15. *Recommended Dietary Allowances,* 8th ed. (Washington, D.C.: National Academy of Sciences, 1974).

quirements are high in growing children whose needs per unit size (in kcal/kg of body weight) are much greater than those of adults. During pregnancy it has been estimated that a woman requires about 80,000 additional kilocalories for a normal 266-day gestation period, or 300 kcal/day. During lactation and breast feeding, these requirements jump to roughly 500 kcal/day. The greater amount of stored fat in women serves as an energy reserve and can provide these extra calories, especially in situations where additional food is not readily available. Disease states may increase or decrease energy requirements. Ordinarily, energy is expended at high rates during infection-induced fevers because of the elevated metabolism, even though this high rate of expenditure is counteracted somewhat by the inactivity associated with illness and fever. Gastrointestinal disorders (such as infantile diarrhea) and intestinal parasites can reduce the ability of the body to absorb nutrient materials, including food energy, vitamins, and minerals. Nitrogen loss, which can deplete protein reserves in the body, also accompanies infection.[16]

Standardized energy requirements by sex and roughly by age (and weight) are presented in table 4.1. These values are appropriate standards for a U. S. or European population where individuals lead a moderately active but not strenuous life. However, these standards would be considerably different for most non-Western populations who would engage in more strenuous work, who would be exposed to environmental conditions markedly dissimilar from the United States or Europe, and who would be most likely to be smaller in size.

Protein from food provides the amino acids necessary to construct many of the body's tissues as well as facilitating a variety of biochemical reactions in the body. However, just "basic" protein is not sufficient to keep the body operating properly.

16. Ibid.

TABLE 4.1

BASAL AND TOTAL DAILY ENERGY
REQUIREMENTS FOR MALES AND FEMALES

Body Weight (kg)	Age (yr.) Males	Age (yr.) Females	Energy Requirements: BMR kcal/day Males	BMR kcal/day Females	Total kcal/day Males	Total kcal/day Females
3	Birth		150	150	350	350
9	10 mos.		500	500	970	970
13	2 yrs.		660	660	1,300	1,300
20	5 yrs.		870	850	1,800	1,750
30	8.5	9.0	1,140	1,060	2,400	2,200
44	13.0	12.5	1,400	1,300	2,800	2,400
54	14.5	17.0	1,550	1,460	2,900	2,100
58	15.5	adult	1,600	1,520	3,000	2,000
70	adult	—	1,780	—	2,700	—

(Adapted from National Academy of Sciences, 1974. **Recommended Dietary Allowances.** Washington, D.C., and World Health Organization, 1973. **Energy and Protein Requirements,** technical report no. 522. Geneva.)

Rather, an appropriate combination of the nine essential amino acids (the "building blocks" of protein) must be present in dietary protein for it to be fully utilized. Requirements range from more than two grams of protein per kilogram of body weight in infants to less than one g per kg in adults. These differences are related to the special needs of growth. Pregnant and lactating women require more protein (up to 50 percent more) than other women because these requirements are also related to growth of the fetus or the breast-feeding infant. Since proteins serve the dual function of providing the basic structural units of the body and of being an energy source, then inadequate food intake of high-energy fats and carbohydrates may force the body to use dietary protein for energy. This case particularly holds when body fat deposits are low and the body has minimal energy stores. As noted above, disease states or any number of stresses can induce nitrogen loss in the urine and hence increase protein requirements.

Other essential nutrients in the diet should include two forms of fatty acids, fat-soluble and water-soluble vitamins, and minerals (both those needed in large amounts and the trace elements). The fat-soluble group comprises vitamins A, D, E, and K. Of these, vitamin D can be synthesized (produced) by the action of ultraviolet radiation on biochemicals in the skin and need not be a part of the diet. The amount of vitamin D produced, however, depends on the amount of available solar radiation and the degree of skin pigmentation. Water-soluble vitamins consist of ascorbic acid (vitamin C), thiamine, riboflavin, niacin, folic acid, choline, pantothenic acid, biotin, and vitamins B_6 and B_{12}. Mineral elements needed in relatively large amounts are sodium, potassium, chorine, calcium, phosphorus, and magnesium. All minerals serve a variety of functions; however, the first three are important to body fluids and the structure of cells, while the second three are important in the growth and composition of bone. There are a large number of trace elements needed in the body. Those required in larger amounts are iodine, copper, iron, zinc, fluorine, and chromium. Equally essential but needed in smaller amounts are manganese, cobalt, selenium, nickel, tin, and several others.

Ecosystem Productivity and Human Food Acquisition

Food practices and dietary intake vary to a large degree throughout the world. Circumpolar and Arctic ecosystems commonly provide large amounts of animal protein and fat but little plant food and low amounts of vitamin C.[17] Moreover, although clinical symptoms of vitamin D deficiency have never been uncovered for Eskimo hunters living above the Arctic Circle, it is likely (in the absence of adequate dietary vitamin D) that insufficient amounts of vitamin D would be synthesized in the skin during the long, dark Arctic winters. Tropical forests on the other hand tend to provide diets that are high in plant foods, both gathered and cultivated, but low in good quality protein and often low in iodine. Arid zones in the tropics (including desert and dry savanna) are characterized by minimal plant and animal diversity and biomass and hence a paucity of food energy. These same areas tend also to be somewhat deficient in several trace elements. Temperate ecosystems (including woodlands and grasslands) usually provide a more varied and hence balanced diet because of rich soils and high plant pro-

17. *Research Programs Constituting U. S. Participation in the International Biological Program,* report no. 4 of the U.S. National Committee for the IBP (Washington, D.C.: National Academy of Sciences, 1971), p. 98.

ductivity. These ecosystems have been most intensively exploited by use of modern agricultural and animal husbandry practices and by extensive use of fossil-fuel energy.

Several examples of ecosystem/dietary interactions and how they influence human nutritional status and general health are discussed below.[18]

Anemia is a multiple-cause condition common throughout the world.[19] One category of anemia is caused by any one of many nutritional deficiencies. These include folic acid and vitamin B_{12}, iron, and even protein deficiencies. Although nutritional deficiency anemias are not exclusively found in the wet tropics, human populations from these parts of the world tend to be especially prone to them. Southeast Asian populations are a case in point. Because much of this area is tropical forest where animal and fish protein is not abundant, both iron and vitamin B_{12} are limiting; hence conditions are especially ripe for nutritional deficiency anemias. Moreover, the relative absence of fresh fruits and vegetables (sources of folic acid) in a diet heavily weighted toward rice is an additional anemia-producing condition.[20]

Protein malnutrition in children, or *kwashiorkor,* is commonly found in the wet tropics and is associated with low intake of animal and plant protein, especially of one or more of the essential amino acids. In the tropical forests of Africa, Central and South America, Southeast Asia, India, and New Guinea, animal products are consumed in small amounts and most of the cultivated starchy plant staples have low-protein content. These include: maize (corn), manioc (cassava), plaintains, bananas, potatoes, yams, and taro. The disease is usually restricted to young children, especially those who have just been weaned from the protein-rich mother's milk and who require substantial amounts of protein for normal growth and development. Moreover, when low-protein intake is associated with low-calorie intake (protein-calorie malnutrition), the minimal protein in the diet will be used as an energy source rather than as an essential protein and amino acid source for the body. As noted in chapter 5, a population will frequently expend large amounts of work energy to maximize the extraction of high-protein foods from the environment.

Iodine-deficiency diseases, such as goiter and cretinism, are found in populations that inhabit inland and mountainous regions. Iodine concentration in soils largely determines its content in plants. Depletion of soil iodine can result from leaching of soil nutrients (as in tropical rain forests and river flood plains) or removal of topsoil through erosion and past movements of glaciers (as in highland zones). The presence of goiter in some populations that live over iodine-deficient soils is further complicated by the presence in the diet of goitrogenic substances, some of which actually inhibit the use of ingested iodine.

As a last example, the trace elements selenium, chromium, and particularly zinc are limiting factors in the diets of many Arab children from the relatively arid zones of the Middle East.[21] In one village in Iran, zinc deficiency was found to be associated with stunted growth. However, the deficiency was produced largely by the con-

18. For a discussion of some other nutritional factors associated with human behavior and genetics, see, Jane H. Underwood, *Biocultural Interactions and Human Variation* (Dubuque, Ia.: Wm. C. Brown Company Publishers, 1975), pp. 22-5.

19. Anemia is a condition in which there are insufficient amounts of red blood cells and constituent hemoglobin.

20. V. Herbert, "Megaloblastic Anemia as a Problem in World Health," *American Journal of Clinical Nutrition* 21 (1968):1115-20.

21. *U.S. Participation in the International Biological Program,* report no. 6 of the U.S. National Committee for the IBP (Washington, D.C.: National Academy of Sciences, 1974), p. 82.

sumption of unleavened bread, which contains a substance that chemically binds zinc and prevents its use by the body. This then is another case of limiting nutrient factors whose availability may be even lessened by dietary interactions.

DISEASE STRESS

Disease is probably one of the most pervading forms of stress experienced by human populations. It also acts as a potent force of natural selection. A state of disease is difficult to define, but in the words of Jacques May it is a "maladjustment" or maladaptation within a given ecological setting, or "that alteration of living tissues that jeopardizes their survival in their environment."[22] A rough classification of diseases can be considered as the following:

1. *Infectious*—A disease produced by bacterial, viral, protozoan, fungal, and parasitic agents.
2. *Nutritional Deficiency*—Disease produced by inadequate, imbalanced, or overabundant nutrient intake.
3. *Environmental Trauma*—Disease produced by any substances that can produce direct or indirect damage to the body.
   a. *physical agents*—environmental pollutants such as aerosol, particulate, corrosive, toxic, or radioactive substances.
   b. *climatic agents*—sunburn, frostbite, windburn, heat exhaustion, hypothermia, hypoxia.
   c. *biological agents*—allergies.
4. *Congenital*—Defects or disease produced by genetic or developmental abnormalities before birth.
5. *Degenerative*—Diseases associated with aging processes, such as various forms of cancer, cardiovascular disease, kidney failure, etc.
6. *Behavioral or Mental*—Abnormalities associated with difficulties or inabilities to operate successfully within a given society.

All disease classification systems are to one degree or another unsatisfactory, and the one above is no exception. For example, despite the separate category "environmental trauma" we know that there are environmental (and genetic) factors involved in the causes of all disease states. Moreover, these categories interact to a great extent; nutritional deficiencies, infectious agents, and environmental pollutants can produce congenital defects, the degenerative aging process can be accelerated by environmental agents (as in lung cancer), and nutritional deficiencies can reduce the effectiveness of the immune response to various infections. Several examples of these interactions can be cited. Congenital malformations (those present at birth) can be produced by a variety of environmental as opposed to genetic factors.[23] The tragic effects of the tranquilizer thalidomide and infectious diseases rubella (German measles) and syphilis on the human embryo are well known from the popular press. In addition to specific nutrient deficiencies (vitamin and protein), there is some evidence that general nutritional deficiency in pregnant women increases the incidence of mental deficiency of those children born whose critical gestation took place during the summer months.[24] The remainder of this sec-

22. Jacques M. May, *The Ecology of Human Disease* (New York: MD Publications, Inc., 1958), p. 1.

23. I. W. Monie, "Influence of the Environment on the Unborn," *California Medicine* 99 (1963):323-7.

24. H. Knobloch and B. Pasamanick, "Seasonal Variation in the Births of the Mentally Deficient," *American Journal of Public Health* 48 (1958):1201-8.

tion on disease stress will focus on infectious diseases at the human population and ecosystem levels.

Infectious Diseases

The maintenance and transmission of infectious diseases in human populations is dependent on contact, either directly between humans or indirectly through an intermediate host or vector. As a result, most infectious diseases are *density-dependent;* that is, the distribution, size, and degree of nucleation (the *density*) of a human population will influence the spread in time and space of a given infectious disease. *Density-independent* factors in the transmission of infection are those environmental factors, such as climate, that are not linked to human population density. In the case of an epidemic, or rapid spread of a disease, the three principal interacting factors are: (1) the density of the human-host population; (2) the biological characteristics of the pathogen (infectious agent); and (3) the density-independent factors.[25] Epidemic diseases cannot survive in populations that are small and widely-dispersed, as in hunting and gathering groups, unless there is an intermediate host to maintain the pathogen. For example, measles (a viral infection) requires a dense population of about 500,000 persons to persist on a year-round basis without constant reintroduction.[26] Malaria and yellow-fever can persist in populations with low density only because of their mode of transmission via the mosquito as intermediate host.

When a new infectious disease is introduced into a "virgin" human population, as occurred following the discovery of the New World by Europeans, the effects are usually devastating. The population will have no genetic or acquired immunity; hence infection and death rates will be high, and there may be successive epidemic waves of the same or different diseases that can decimate the population. During this period there is often an intense selection process going on in both the human host and pathogen populations. In the human population those who survive the epidemic have developed some degree of immunity or have been selected for an ability to resist the disease. In the pathogen population there is a corresponding selection, but for reduced severity of effect that will promote survival of the pathogen; for if all hosts die, so will the pathogen. This general pattern of interaction will often lead to a state where a previously serious threat to human survival becomes merely an inconvenience. In this way a *parasitic* relationship between human host and pathogen can evolve into a more viable form of ecological *symbiosis,* where both species can survive. This reduction of detrimental effects usually occurs in species that have been in contact for long periods under relatively stable ecosystem conditions.[27] Therefore the continual outbreaks of epidemics in human populations, in addition to the high population density factor, is probably the result of continual human movement and cultural and ecological change.

Sociocultural factors also play an important role in the destructive effects of epidemics. A recent example of this has been documented for the Yanomama Indians of Brazil and Venezuela. Having only recent contacts with Europeans, this population of tropical forest dwellers of the Amazon Basin was struck by a serious measles

25. K. E. F. Watt, *Principles of Environmental Science* (New York: McGraw-Hill Book Co., 1973), p. 229.

26. F. Fenner, "The Effects of Changing Social Organization on the Infectious Diseases of Man," in *The Impact of Civilization on the Biology of Man,* ed. S. V. Boyden (Toronto: University of Toronto Press, 1970), pp. 48-68.

27. E. P. Odum, *Fundamentals of Ecology,* 3d ed. (Philadelphia: W. B. Saunders, 1971), p. 220.

epidemic in 1968.[28] Although medical personnel attempted to contain the disease by use of measles vaccine, the disease reached epidemic proportions in a number of villages where nearly all inhabitants at all ages were infected. In addition to the complicating factors of associated respiratory infections and dehydration from fever in the tropics, normal village life was totally disrupted, and care for the ill was nonexistent. Nursing children who were sick could not be tended by sick mothers, the sick were abandoned, and the general psychological effect produced by this epidemic was either fatalism or panic. It is possible to generalize from this one example by suggesting that in most cases of widespread illness, the population is faced not only with the direct effects of the disease but also with secondary infections and a breakdown of the basic life-support system of the population.

Although epidemic diseases can literally decimate a population (as with plague in Europe during the Middle Ages and smallpox in the New World during the sixteenth century), *endemic,* or habitually present diseases, have probably played a more important role in the evolution of human populations throughout their history. Even today, gastrointestinal diseases (i.e., gastroenteritis), respiratory infections, parasitic infestations, and diseases caused by a wide spectrum of pathogens are major killers among non-technologically-advanced peoples around the world. Ecological relationships, of course, are of primary importance in the incidence and prominence of these diseases. For example, Dunn astutely pointed out that the biotic diversity variation found among ecosystems also holds for pathogens that affect humans.[29] There is a generalization that in "complex" ecosystems, such as a tropical rainforest, the diversity index is high, and although there are many species and niches, there are relatively few individuals per species. A "complex" ecosystem can be contrasted with a "simple" ecosystem, such as a desert or Arctic tundra, where the diversity index is low, and few species and niches are represented by relatively large numbers of individuals per species. Dunn compared African Bushmen and Australian Aborigines, both inhabiting desert ("simple") ecosystems, with Malayan Semang and African Pygmies who live in tropical rain forests ("complex") ecosystems. With respect to certain protozoan and parasitic diseases, the tropical rain forest dwellers were carrying an infectious load of from two- to seven-times the number of species carried by the desert dwellers! However, it is important to note that the few species of pathogens carried by the "simple" ecosystem dwellers may be a serious threat to their health because of the greater intensity of infection associated with large numbers of individuals within a pathogen species.

Perhaps one of the most serious health hazards to human populations results from the interaction of gastrointestinal infections, intestinal parasitic infestation, and protein and/or calorie malnutrition. Infants and young children are especially threatened by this combination. Because infants and children have a small body size and not much

28. W. R. Centerwall, "A Recent Experience with Measles in a 'Virgin-Soil' Population," in *Biomedical Challenges Presented by the American Indian,* Pan American Health Organization, scientific publication no. 165 (Washington, D.C.: World Health Organization, 1968), pp. 77-81; J. V. Neel, "Genetic Aspects of the Ecology of Disease in the American Indian," in *The Ongoing Evolution of Latin American Populations,* ed. F. M. Salzano (Springfield, Ill.: C. C. Thomas, 1971), pp. 561-90.

29. F. L. Dunn, "Epidemiological Factors: Health and Disease in Hunter-Gatherers," in *Man the Hunter,* ed. R. B. Lee and I. DeVore (Chicago: Aldine Publishing Company, 1968), pp. 221-8; "biotic diversity variation" is a measure of the numbers and varieties of species that make up the biotic community of an ecosystem.

body fat reserves, diarrhea can cause a rapid water loss (dehydration), body weight can drop equally rapidly, and susceptibility to other infections (especially respiratory diseases) increases. This general state of poor health has been documented quite clearly for infants in Gambia, West Africa. It is remarkable that some infants survive successive bouts of gastroenteritis, malaria, whooping cough, and kwashiorkor, all within the first two years of life.[30]

DEMOGRAPHIC STRESS

It is difficult to deal with the notion of what constitutes human population pressure or stress, and how controls are exercised to reduce this stress to maintain human populations at optimal levels. Such problems are linked to our inability to identify what an optimal human population size is and what the so-called "carrying capacity" of an ecosystem is for human-population numbers.[31] Carrying capacity is an unsatisfactory concept to deal with for a number of reasons: First, it has a time-depth element, based on limiting and tolerance factors, that may be seasonal, annual, or may fluctuate over longer time periods. An example can be taken from the case study of savanna pastoralists from chapter 5. For these groups population numbers are probably adjusted to the minimal resources available during the dry season, although longer period droughts (each ten to twenty years) may also influence population size and distribution at any given time. Second, a relatively minor social or technological change can rapidly enable an ecosystem to support larger numbers of people. An example can be taken from the case study on the Altiplano Indians of Peru, where increased or decreased trade with populations at lower altitudes can correspondingly produce conditions supporting greater or fewer numbers of people.[32] Third, if an ecosystem is to be maintained in a relatively balanced state, then the human carrying capacity will depend on an established equilibrium struck between human numbers and populations of other species, all of which constitute the biotic community. At the expense of being accused of circular reasoning, one might simply identify carrying capacity (in what appears to be a balanced ecosystem) as a state of human population numbers likely to be the optimal or adaptive state at a given time. Any major change in human population numbers is then almost certain to be disruptive to both the human population and the ecosystem and requires ecosystem adjustments to reestablish equilibrium.

Human population controls ultimately can be viewed according to the simplified formula:

$$P_1 = P_0 + \text{Births} - \text{Deaths} \pm \text{Migration}$$

where a population at a given time (P_1) is the result of its size at an earlier time (P_0) with additions through births and in-migration and losses through deaths and out-migration. Birth rates are a reflection of human fertility patterns, death rates, repre-

30. A. M. Thomson, "The Development of Young Children in a West African Village: A Study in Human Ecology," in *Human Ecology in the Tropics*, ed. J. P. Garlick and R. W. J. Keay (London: Pergamon Press, 1970) pp. 99-112.

31. The usefulness of this concept, which refers to the numbers of people that an environment can support, has been repeatedly challenged. For a recent discussion of the limitations of the carrying capacity concept, see, B. Hayden, "The Carrying Capacity Dilemma: An Alternative Approach," in *Population Studies in Archaeology and Biological Anthropology*: A Symposium, ed. A. C. Swedlund, *American Antiquity* 40 (1975), no. 2 part 2:11-21.

32. P. T. Baker, "Discussion," in *Population Studies in Archaeology and Biological Anthropology: A Symposium*, ed. A. C. Swedlund, *American Antiquity* 40 (1975), no. 2 part 2:44-5.

sent Malthusian controls, and in- and out-migration serve to augment the controls of differential fertility and mortality. Figure 4.4 illustrates the effects of relaxation of mortality control, a subsequent response through fertility control mechanisms, and the net effect on size of an hypothetical human population. When human populations are in a state of ecological balance, birth and death rates are roughly equal and population size remains constant through time (left side of figure 4.4). If some outside disturbance, such as the introduction of modern medicine to a non-Western population, produces a rapid decline in death rate (center of figure 4.4) while birth rate remains high, the population will dramatically increase in numbers. Under ideal control conditions birth rate will begin to decline, population growth will stop, and a new equilibrium will have been established, although at the cost of the ecosystem now having to support a larger population (right side of figure 4.4). This pattern may produce an equilibrium state for birth and death rates and consequent numerical population stability, yet may in fact produce a state of disequilibrium with respect to the population's ability to maintain its vital life-support system. Since a population's life-support system is dependent on some degree of ecosystem balance, it is both impractical and unrealistic to attempt to consider human population controls apart from the ecological context.

A schematic representation of increased population control through mortality is shown in figure 4.5. When death rates increase for any number of reasons, population numbers will decline. Birth rates may increase to compensate for population loss-

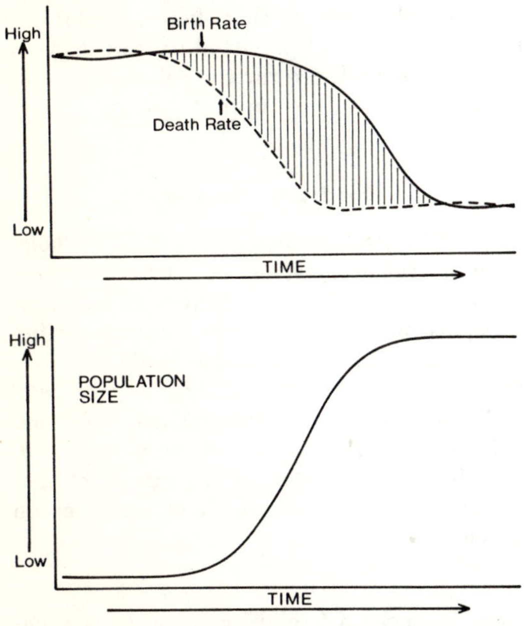

FIGURE 4.4. Population increase through declines in mortality followed later by declines in fertility.

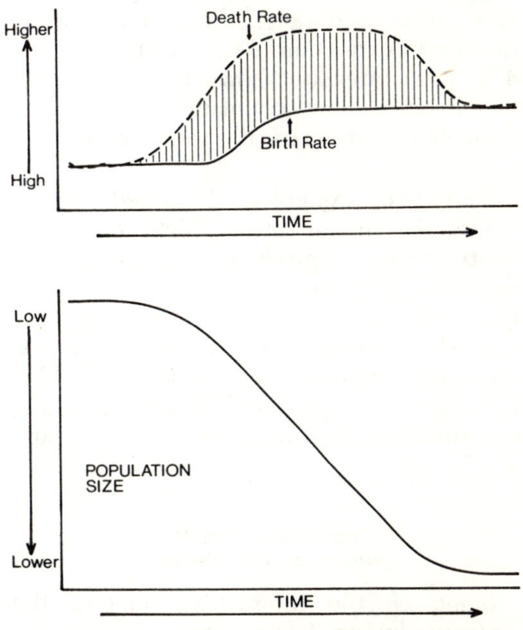

FIGURE 4.5. Population decrease through rises in mortality followed later by only slight rises in fertility.

es; although if mortality is widespread and crosscuts all age ranges, it is just as likely that fertility will decline. In any case, where death rates rise precipitously (as shown in figure 4.5), an increase in birth rate can only partially compensate for the losses. This results from the fact that most non-Western populations are close to the maximal fertility (*completed fertility* of between five and ten live births) and further increases in birth rate can only be slight.[33] Another factor that influences the configuration of these curves is the age distribution of death rates. If most deaths are among those who are past the reproductive years, the effect on population decline is not as great as if most deaths are among children or young adults. If the latter were the case, the population's ability to recover would be limited, since overall birth rates would then be reduced. In general, if death rates remain above birth rates (in the absence of in-migration), the population will decline to extinction. At the right of figure 4.5, death rates are shown to drop off, establishing a new population equilibrium with its size dramatically reduced in numbers.

The two examples described above, where the interaction of birth rates and death rates can produce an increase or a decrease in population size, are extreme cases and represent macrochanges in these variables. Throughout most of human history, population fluctuations are more likely to have been produced by microchanges, or smaller-scale variations, in birth and death rates.

Population, Climate, Nutrition, and Disease

Some of the interactions among the stress-producing factors discussed in this chapter are dealt with in the case studies that follow. The principle stresses imposed on Eskimo populations are probably climatic and nutritional in nature, although the problems of Westernization are certain to introduce additional stresses. Among the savanna pastoralists, unpredictable rainfall, human and cattle disease, competition for pasturage, and the constant threat of starvation during drought years are major sources of population stress. Highland Quechua Indians of the Peruvian altiplano must deal with hypoxia, cold, altitude-induced or -related diseases, limited energy resources, and population pressures for scarce land resources. New Guinea natives are faced with nutritional stresses due to low protein availability, associated nutritional diseases, diseases related to a tropical rain forest ecosystem, and some population pressure for land resources. Each population, however, has developed a unique life-support system to deal with these and other stresses and environmental problems with which they are continually confronted. The fact that these populations have survived and persisted up to the present time is evidence for the relative success of each of these life support systems.

For Further Reading

Baker, P. T. and Weiner, J. S., eds., *The Biology of Human Adaptability*. Oxford: Clarendon Press, 1967. A collection of overview papers on the environmental biology of human populations around the world.

Folk, G. E., Jr. *Textbook of Environmental Physiology*, 2d ed. Philadelphia: Lea and Febiger, 1974. An introductory but comprehensive treatment of the effects of biological rhythms, climate, and radiation on man.

33. "Completed fertility" is taken, for a population, as the average number of live births per woman during her reproductive lifetime (usually estimated as between fifteen and forty-five years of age).

Food: Readings from Scientific American. San Francisco: W. H. Freeman and Company, 1973. A collection of twenty-eight excellent and readable papers from past issues of Scientific American.

Swedlund, A. C., ed. *Population Studies in Archaeology and Biological Anthropology: A Symposium, American Antiquity* 40, 2 (1975). This special journal issue explores many interesting demographic problems.

Slonim, N. B., ed. *Environmental Physiology.* St. Louis: C. V. Mosby Company, 1974. This is a more comprehensive collection of review chapters dealing with a variety of environmental influences on humans.

Bibliography

Boyden, S. V., ed. 1970. *The Impact of Civilization on the Biology of Man.* Toronto: University of Toronto Press.

Damon, A., ed. 1975. *Physiological Anthropology.* New York: Oxford University Press.

Le Riche, W. H. and Milner, J. 1971. *Epidemiology as Medical Ecology.* Edinburgh: Churchill Livingstone.

5 Ecological Case Studies of Human Populations

The case studies of which this chapter is comprised deal with four human populations that employ quite diverse life-support systems: *Arctic Eskimos, East African Savanna Pastoralists, Peruvian Altiplano Indians,* and *New Guinea Highlanders.* Each was chosen because the particular population has been studied in some depth within an ecological framework. Two of the populations, *Peruvian Altiplano Indians* and *New Guinea Highlanders,* were investigated under field conditions by the authors. For clarification of some of the materials presented in the case studies, the reader may wish to refer to a few of the principles discussed in the preceding chapters.

ARCTIC ESKIMOS

Eskimo hunters are distributed along coastal and some inland regions of Alaska, Canada, and Greenland. Probably no other native populations anywhere in the world today have maintained such a high degree of cultural homogeneity while being so widely dispersed as the Eskimo. Although ethnographic studies have surveyed many populations throughout this broad geographic zone, few studies have focused on the human ecology or biology of Eskimos.

The case study that follows is based on information from three populations that have been investigated within these frameworks. First, on the north coast of Alaska, the settlements of Wainwright and Barrow were studied by Frederick Milan and his associates as a part of the International Biological Program (IBP) research in human adaptability. Second, Eskimo populations from Igloolik and Hall Beach on the northwest coast of the Melville Peninsula were studied as part of the Canadian IBP. Finally, Eskimos from Lake Harbour on the south coast of Baffin Island were investigated by William Kemp as a model of energy-flow relationships. Figure 5.1 locates these three populations on a map of the North American Arctic.

The Tundra Ecosystem

The Arctic tundra is characterized by soils with a permafrost base, low species diversity and density, and a short plant growing season that varies between two and three months. Most of the conditions in the tundra zone are produced by a combination of low air temperatures, precipitation, evaporation, and radiation.

During the short cool summers temperatures range between freezing and about 10°C (50°F). With the thawing of the light snow cover, the relatively few species of lichens, low shrubs, flowering herbs, sedges, and grasses rapidly mature to cover the wet landscape with a dense carpet of vegetation. At this time the permafrost will thaw only to a depth of less than one meter, limiting plant root growth and contributing to poor water drainage. This in turn produces a series of raised meadows, ponds, marshes, and thaw lakes that remain throughout the summer months. Travel during this period, especially in the late spring, is difficult because of the marshy conditions, and it is at this time that the landscape is most susceptible to damage by vehicular movement.

The winters in the Arctic are long and cold with intense wind chill, producing conditions of potentially acute cold stress imposed on the human population. For example, in Barrow, Alaska, the average January temperature is -26°C (-16°F), with periodic cold spells as low as -40°C (-40°F). Much of the surface water freed by the summer thaw is again trapped by freezing during the winter months and augmented by snowfalls that usually accumulate to well under a meter. Annual precipitation throughout the Arctic tundra zone is generally less than 25 cm, a characteristic of arid or semiarid climates. During the early fall, the onset of winter is signaled by the formation of ice slush along the coast line and then sea ice which may extend several miles beyond the coast.[1] Sea ice usually reaches its maximum extent by late March, at which point movement and transportation is easiest with Eskimo technology.

A simplified model of the wet tundra from research conducted at Barrow by IBP ecologists is illustrated in figure 5.2. This is a model of the basic biotic components of the tundra ecosystem and how energy and materials flow through the system. Nutrient pools are released during the spring thaw and used by the growing vegetation, or primary producers. The vegetation is fed on

1. R. K. Nelson, *Hunters of the Northern Ice* (Chicago: University of Chicago Press, 1969).

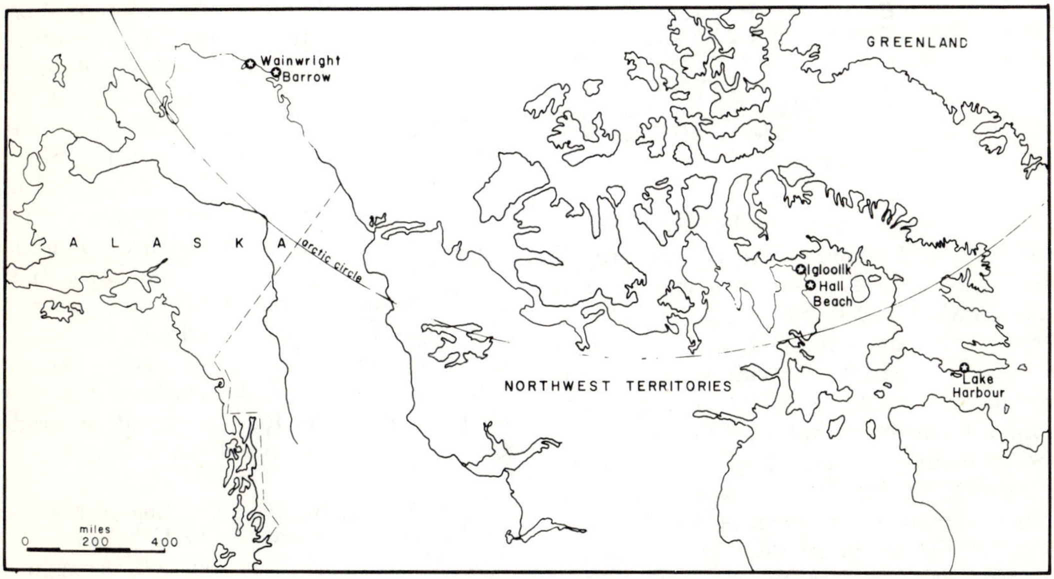

FIGURE 5.1. Map of the North American Arctic.

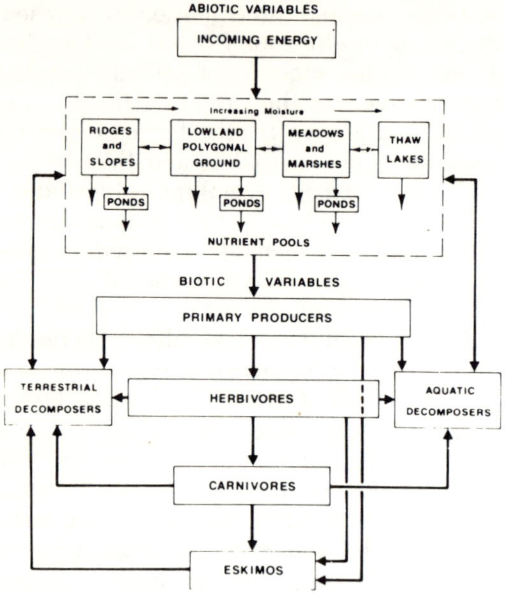

FIGURE 5.2. Terrestrial energy and nutrient flows in the Arctic tundra of northern Alaska. (Adapted from *Research Programs Constituting U.S. Participation in the International Biological Program.* 1971. Report no. 4 of the U.S. National Committee for the International Biological Program. Washington, D.C.: National Academy of Sciences, p. 33.)

by herbivores, such as the brown lemming and caribou. Lemmings in turn are preyed on by the carnivorous fox, jaeger, and snowy owl. The Eskimo component of this terrestrial model would be identified as a higher-order carnivore, with some qualification, since Eskimos utilize food not only from the carnivore but also the herbivore and primary-producer trophic levels. In addition to the food resources of the tundra, Eskimos exploit the marine food chains from the sea ice to open-sea environments, perhaps even rivaling the decomposers in the diversity of foods consumed.

Alaskan and Canadian Eskimos Today

Eskimos are experiencing changes in conjunction with increasing acculturation and life-style modification, the urbanization trend, and environmental degradation, all of which can produce a state of disequilibrium among these peoples. The traditional Eskimo life of hunting and fishing in the tundra and sea-ice environments by use of harpoon, kayak, and dogsled is rapidly undergoing change. The harpoon has long since been replaced by the rifle, the kayak by the motor-powered boat, and the dogsled, within the past ten or fifteen years, by the snowmobile.[2] Each new "improved" mode of exploiting the environment has required a greater involvement in channels of monetary exchange, such that the Eskimo is increasingly compelled to become either a part-time wage earner or active tradesman. As an example, although the snowmobile expands the hunting zone of exploitation, the support of this motorized vehicle demands substantial wage income.

Despite this greater involvement in a money economy and incorporation of Western technology, the Eskimo still devotes a large proportion of his time to hunting activities. In Alaska and Canada, seasonal migrations of caribou and waterfowl provide huntable terrestrial game. Marine mammals, such as a variety of seals, the walrus, and the beluga whale, are hunted on the sea ice and the open sea; fish are abundant, too. The hunting of sea mammals for nearly all coastal Eskimo populations provides the bulk of food from hunting, fishing, or gathering activities. Kemp, for example, found that for the Lake Harbour Eskimos, 83 percent of the food acquired by traditional means constituted sea mammals.[3]

2. E. S. Hall, Jr., "The 'Iron Dog' in Northern Alaska," *Anthropologica* 13 (1971):237-54.

3. W. B. Kemp, "The Flow of Energy in a Hunting Society," *Scientific American* 225, no. 3 (1971):104-15.

Figure 5.3 illustrates the annual cycle of traditional subsistence activities in one Eskimo camp near Igloolik. This figure well represents the seasonally determined availability of food animals and the patterns of work. It does not, however, represent the complexity of the patterns of movement and settlement activity which are the key attributes of the resource-management aspect of the Eskimo life-support system. Eskimo local populations or village settlements are widely scattered, often ten to thirty miles apart. Within local populations or camps, Eskimos respond to seasonal variations in food availability by modifiying population distribution in three basic ways: (1) shoreline or coastal dispersal (spring); (2) inland concentration (summer); and (3) shoreline concentration (fall and winter). With the onset of spring and longer days, the population begins to disperse into more widely scattered camps. This is done to improve hunting of the more visible and concentrated seals on the now fragile sea ice, and to hunt the less abundant and more mobile sea and terrestrial mammals. By midsummer, groups have moved inland to more concentrated settlements to intercept the regular caribou herds. Large camps are established at the heads of valleys accessible to higher-altitude areas. During this period, other small terrestrial game is taken, along with fish and small amounts of plant foods. The Eskimos return to sometimes permanent settlements along the coast at the close of summer and may depend on stored food until the establishment of the smooth sea ice in early winter. Winter activities are focused on the shore or on the sea ice in protected bays. It is on the temporary coastal plain created by the sea ice where the principal food source, the seal, is hunted. The winter season is clearly the most precarious, since any interruption in subsistence activities of more than a week may bring the population to the brink of starvation.

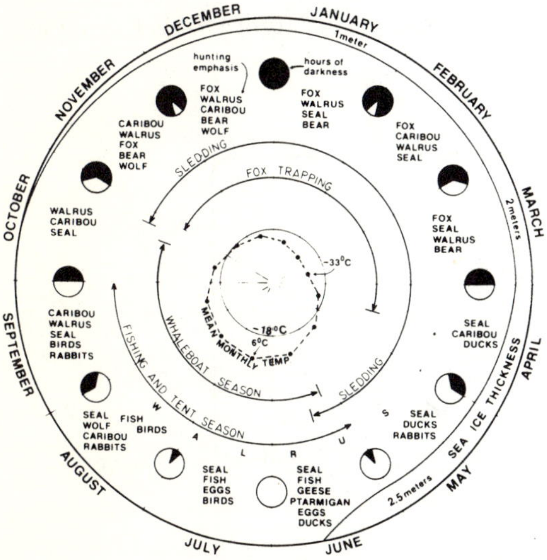

FIGURE 5.3. Annual cycle of a Canadian Eskimo settlement. (Adapted from Godin, G.J. 1972. "A Study of the Energy Expenditure of a Small Eskimo Population." Physiology section report, IBP Human Adaptability Project, Igloolik, N.W.T. Toronto: University of Toronto Press.)

The population-control aspect of the Eskimo life-support system is of particular interest because of the nature and intensity of population-controlling responses reported for these groups. For example, the rate of interpopulation movement facilitated by local exogamy (out-breeding) can readily respond to local habitat quality and population pressure. Marriage patterns establish a network of relationships that can function to add flexibility to resource-management strategies by facilitating short- and long-term visits and temporary shifts of residence. Eskimo sexual behavior is reported to be free and imaginative within a circle of established relationships, a factor which should maximize reproductive performance. On the other hand male and female infanticide is (or was) regularly practiced on those infants born during periods of intense life-support stress. Other population controls include the sporadic practice of selective female infanticide, polygynous marriages, the high risk of accidental death for adult men, and the widespread practice of gerontocide (abandonment of the elderly).

The dwellings, clothing, and other aspects of Eskimo material culture and technology have undergone considerable change over the years. Sod houses and skin tents are rare, having been replaced by frame houses provided or subsidized by the Canadian and U.S. governments. Dwellings still are heated in many areas by oil lamp (using sea-mammal fat), although kerosine is being increasingly used, especially by Eskimo groups close to trading or larger population centers. The cost of such fuels is great, however, about double the price asked outside of Alaska and the Canadian Northwest Territories. The traditional skin clothing has been replaced by tailored Western clothing in many areas of the Arctic, but there is still some reliance on skin clothing in other areas.

Energetics of Eskimo Populations

Figure 5.4 is an energy-flow diagram derived from the work of Kemp in his studies of two Eskimo families from Lake Harbour.[4] Among these Eskimos, the energy consumed is derived from three primary sources: food, fuel, and ammunition. Food is either purchased or hunted and gathered; ammunition is purchased; and fuel is derived from sea-mammal fat or purchased as kerosine and gasoline.

The upper part of figure 5.4 is the money-

4. Ibid.

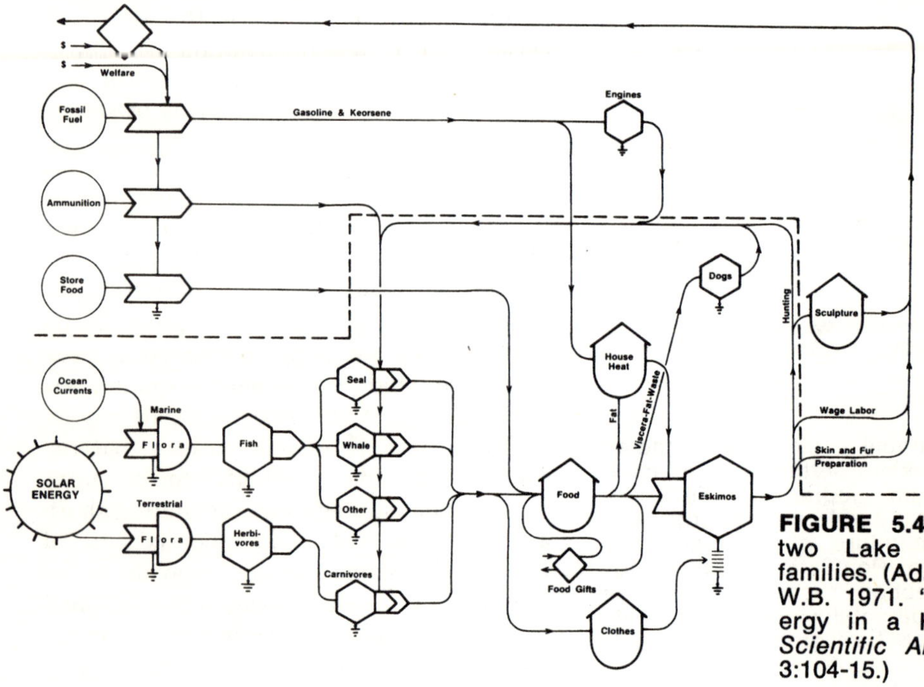

FIGURE 5.4. Energy flow in two Lake Harbour Eskimo families. (Adapted from Kemp, W.B. 1971. "The Flow of Energy in a Hunting Society." *Scientific American* 224, no. 3:104-15.)

economy component of Eskimo subsistence. Two principal outputs within the "money" component are wage labor and the manufacture of sculptured Eskimo art objects. A third output (in terms of energy expenditure by activity) is the preparation of animal skins and furs for sale through Western markets. Funds received from these activities are augmented by government welfare payments that amount to nearly $7,000 per year for the two families that were studied at Lake Harbour by Kemp. The principal inputs in the form of purchases are store food, ammunition, and fossil fuels, although funds are also needed to maintain snowmobiles and marine engines and to purchase other goods.

Despite the joys of hunting and the independence of the traditional Eskimo way of life, it is clear why the Eskimo has incorporated Western technological innovations and has become more involved in a money economy. Figure 5.5 summarizes the energetics of food consumption (input) and energy expenditure by activity (output) for the two Lake Harbour families. Hunting with motor-powered boats and snowmobiles requires the expenditure of 1.75 million kcal while providing 5.29 million kcal of food. The energetic efficiency of this pattern of food acquisition is 5.29/1.75 = 3.0, or a three-fold return. On the other hand the activities associated with a money economy (mining and carving sculptures, wage work, and skin and fur preparation) require only an expenditure of 1.12 million kcal while providing 7.55 million kcal of food. The energetic efficiency of this pattern is 7.55/1.12 = 6.7, or a nearly seven-fold return. This example is, of course, a gross oversimplification of the food/activity energy budget, but it does serve to illustrate the efficiency (from the Eskimo standpoint) of the system. It also explains the apparent abandonment of the most stringent population-control measures and the expanding numbers of Eskimos today.

What has not been considered in these calculations is the impact of changing patterns of Eskimo subsistence on their social system, their health, and their ecosystem.

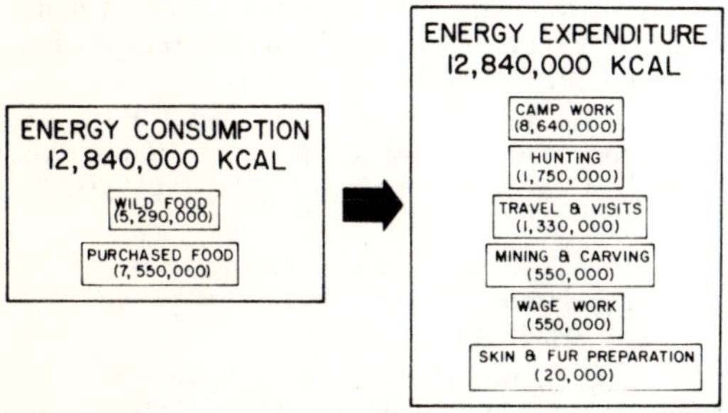

FIGURE 5.5. Food energy consumption and energy expenditure of two Lake Harbour Eskimo families. (Data from Kemp, W.B. 1971. "The Flow of Energy in a Hunting Society." *Scientific American* 224, no. 3:104-15.)

Health and Biology of Eskimos

The Wainwright, Alaska population is basically a healthy one, but certain chronic disorders are common. These include middle-ear infections and pulmonary tuberculosis (TB). Nearly half of this population had a history of TB in the late 1960s, although 82 percent of this proportion were adults. Since very few of the youth have the disease (originally introduced by European contacts), the U.S. Public Health Service programs apparently are successful. As a result of changing nutritional patterns, dental health has declined over the past twenty or so years. The average number of decayed, missing, or filled permanent teeth in Wainwright Eskimos was eight times greater than in the less acculturated Canadian Igloolik Eskimos.[5] Studies of work capacity have shown that Eskimos are more physically fit than a comparable U.S. or Candian white population, and they maintain this fitness into old age.[6]

Eskimo populations in both Canada and Alaska are rapidly increasing in numbers. This results from the joint effects of a high fertility rate and a rapidly declining mortality rate (especially among infants) because of improved health-care facilities. For example, in Wainwright, Eskimo women show an average completed fertility of between nine and ten live births. When combined with the relatively low mortality over the past decade, the population is increasing at a rate of about 3 percent per year.[7] In addition to increasing in numbers, Eskimos are also increasing in size. The secular trend toward increased body size at all ages observed among many other populations is also being expressed by Eskimos. Today, Eskimo adults are an average of two to three inches taller and several pounds heavier than their counterparts of the late 1800s.[8] These increases leading to a greater biomass of Eskimo populations may have serious effects on the carrying capacity of the tundra coastal environment.

Eskimos throughout the Arctic have one of the most specialized high-protein and low-carbohydrate diets anywhere in the world. Animal protein is still consumed as a large proportion of the diet despite the increased carbohydrate intake (sugar, flour, etc.) from purchased foods. At Lake Harbour, Kemp recorded caloric intakes of 44 percent protein, 33 percent fat, and 23 percent carbohydrate. Eskimos may even have a unique metabolic ability to digest such large amounts of protein.[9] A condition of lessened bone mineralization (osteoporosis) in older adults that contributes to a greater risk of bone fractures may be linked somehow to a high-protein diet.[10] Animal protein is high in phosphorus, and the Eskimo diet is usually low in calcium (a principal mineral in bone). Hence the normal processes of bone loss associated with aging may be accelerated by this imbalance in calcium-phosphorus intake. Alternately, it has been observed that the acidic effect of a high-animal-protein diet produces an increase in calcium excretion. If either of these relationships is valid it will be the only nutrition-associated problem observed for Eskimos on a traditional diet.

In contrast to earlier observational studies of skin color in Eskimos, Wainwright Eskimos have levels of skin pigmentation that are quite similar to whites or Europeans. This is particularly interesting, since population variations in skin pigmentation are believed related to the ultraviolet radiation-induced production of vitamin D in the skin. However, since vitamin D is relatively abundant in the Eskimo diet (from fish oils) and the Arctic is characterized by annual extremes of available sunlight, the need for vitamin D production in the skin may be reduced. Thus Eskimo skin color may be more of an adaptive response to other factors associated with Arctic varia-

5. J. T. Mayhall, "The Effect of Culture Change upon the Eskimo Dentition," *Arctic Anthropology* 7, no. 1 (1970):117-21.

6. D. W. Rennie et al., "Physical Fitness and Respiratory Function of Eskimos of Wainwright, Alaska," *Arctic Anthropology* 7, no. 1 (1970):73-82.

7. F. A. Milan, "The Demography of an Alaskan Eskimo Village," *Arctic Anthropology* 7, no. 1 (1970):26-43.

8. P. L. Jamison, "Growth of Wainwright Eskimos: Stature and Weight," *Arctic Anthropology* 7, no. 1 (1970):86-94.

9. H. H. Draper, "Aspects of Adaptation to the Native Eskimo Diet: Energy Metabolism," *American Journal of Physical Anthropology* 41 (1974): 475.

10. R. B. Mazess and W. Mather, "Bone Mineral Content of North Alaskan Eskimos," *American Journal of Clinical Nutrition* 27 (1974):916-25.

tions in solar radiation rather than vitamin D production alone.

Numerous laboratory studies of Eskimo responses to cold stresses have been conducted, but little work has been done to define "thermal environment" in full Eskimo garb. As a result, most of our conclusions on biological adaptation to cold are tentative.[11] There are, however, conclusions that can be drawn from the extensive literature on the topic. First, despite the commonly held belief that the Eskimo face is engineered to combat cold, there is no experimental evidence to support this contention. Second, the basal metabolic rate is higher than most other populations and appears to give the Eskimo a slight advantage during chilling of the whole body. Whether this is produced by the Eskimo's high protein intake or some other cause has not yet been established. Third, Eskimo hand and finger temperatures are consistently warmer than in other populations when tested in the cold. This should be to the Eskimo's advantage by his being able to maintain adequate manual dexterity while engaged in hunting or other activities during the winter months.

Among Eskimos today there is less sensitivity to the balance of nature within their world view and ritual system. As a result, the internalization of Western values and use of Western technology can potentially lead to ecological imbalance within the tundra system. Increased use of snowmobiles will both damage the landscape and extend the ability to kill game. A greater Eskimo biomass resulting from population increases and secular growth advances can only aggravate these ecosystem perturbations. Conversely, the reduced emphasis on hunting and greater involvement in craft manufacture and wage labor will reduce the exploitation of wildlife and contribute in a small way to the maintenance of a balanced tundra ecosystem. Finally, the Eskimo is now paying a price for Westernlization in experiencing declines in some aspects of general health. The adaptation to a high protein/fat diet and high activity way of life that once served to protect the Eskimo from heart disease may be breaking down since cardiovascular disease among the group seems to be on the rise. Tooth decay and certain infectious diseases are also becoming new health problems.

Compared with the traditional self-sufficient Eskimo pattern of subsistence, the pattern that exists today is substantially more complex and unstable. William Kemp feels that a small corps of hunters will continue to exploit the Arctic for some time to come. This will, of course, depend on the ability of the Arctic ecosystem to provide the resources to support such a hunting system.

EAST AFRICAN SAVANNA PASTORALISTS

Savanna, or tropical grassland, ecosystems stretch in in a wide belt across subsaharan Africa north of the equator, in the region of East Africa around Lake Victoria, and in an even wider belt across southern Africa. There is no other continent in the world with as great an expanse of savanna lands as Africa, and no other continent with as large a native population dependent on the savanna at a subsistence level of economy.

This case study focuses on several savanna-dwelling populations from Uganda and Kenya that maintain a traditional nomadic existence tending livestock and, for some groups, periodically growing cultivated crops. These populations include the Turk-

11. For an overview of Eskimo cold adaptation, see, M. A. Little and A. T. Steegmann, Jr., "Acclimatization and Adaptation: Responses to Cold," in *Environment, Origins, and Population, vol. 3, Handbook of North American Indians*, ed. F. S. Hulse (Washington, D.C.: Smithsonian Institution, Bureau of American Ethnology, 1976).

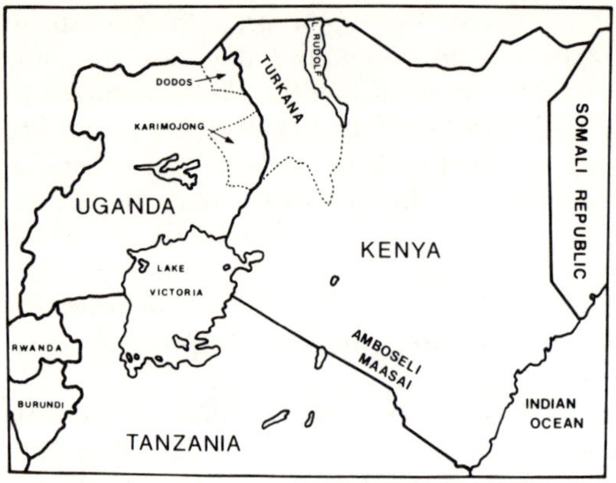

FIGURE 5.6. Map of East Africa.

ana studied by Neville Dyson-Hudson, the Ilkisongo Maasai of the Amboseli game reserve studied by David Western, the Karimojong jointly investigated by Rada and Neville Dyson-Hudson, and the Dodos, a tribe that lives north of the Karimojong studied by W. Deshler. These specific groups were chosen because each was investigated within an ecological framework by scientists with quite different backgrounds; the Dyson-Hudsons are anthropologists, Western is an ecologist, and Deshler is a geographer. These human groups are located on a map of East Africa in figure 5.6.

The Savanna Ecosystem

It is really inappropriate to refer to a *single* savanna ecosystem because there are a variety of subsystems now being defined for East Africa. In general, the term savanna refers to a vegetational zone consisting of combinations of scattered trees and low shrubs within a tall growth of grass. Pratt and others defined five savanna land ecological zones, from humid to very arid, on the basis of climatic characteristics.[12] The humid savanna can support forests and derived grasslands and bushlands, with some economic potential for forestry or intensive agriculture. There is an assumption that if permitted to achieve a climax vegetation state, the humid savanna would convert largely into woodland. However, savanna grasslands are generally classified as such only if the tree canopy (the leafy portions of the trees) covers less than 20 percent of the surface area of the land. The remaining zones from dry subhumid to very arid support increasingly less vegetation, become less appropriate for agriculture, and require more and more land to support a given number of wildlife and livestock. In the very arid zone, grazing land is of very low potential, and the vegetation may be confined only to localized areas.

Despite the importance of climate, particularly rainfall patterns, there are many other factors that influence the ultimate vegetational characteristics of these tropical grasslands. For example, soil type, topographic relief of the land (terrain), and relative presence of groundwater (belowground water reserves) each affect the primary productivity of the vegetation. Secondary effects are produced by wildlife that graze and browse on savanna vegetation, particularly large herbivores, such as the elephant, giraffe, zebra, eland, rhino, gazelle, and African buffalo. Tertiary effects are, of course, the products of human activity and include grazing of livestock, shifting cultivation (where feasible), and burning. It is safe to say that there are few areas of the vast expanse of savanna lands that do not bear the mark of human activities.

Climatic characteristics of the East African savanna are seasonal with one or two rainy periods, but the pattern of a single dry and a single wet season predominates.

12. D. J. Pratt, P. J. Greenway, and M. D. Gwynne, "A Classification of East African Rangeland, with an Appendix of Terminology," *Journal of Applied Ecology* 3 (1966):369-82.

Being within the equatorial zone, seasonal temperature variation is not great, but temperatures do tend to be warmer during the dry season when evaporative loss of standing water and water from the soil is high. Rainfall patterns are quite variable from place to place and from year to year, and the ability to predict the onset of the wet season or the amount of rainfall in any given year is poor. Annual rainfall may range in different zones from less than 10 inches (25 cm) to more than 50 inches (125 cm). The Dyson-Hudsons give an example of this erratic rainfall pattern by noting that in the mid-1950s, two weather stations only eight miles apart within the Karimojong tribal area recorded annual rainfalls of about 20 and 42 inches, respectively![13] In addition to limitations imposed on the life of the savanna by scarce water resources, high temperatures pose other problems to wildlife, livestock, and humans alike in the form of heat stress. Heat stress in turn is aggravated by limited water supplies. The relatively humid savanna can be characterized by its abundant wildlife, high primary productivity of plants, rich grazing lands for livestock, and ability to support cultivation. The arid savanna is a marginal, hostile environment that is subject to drought, has minimal vegetation, and requires nearly constant movement of humans and their livestock in search of water and forage.

Of the several human groups discussed in this case study, the Turkana (who live west of Lake Rudolf in Kenya) inhabit the most arid, marginal lands. Rainfall is commonly less than 10 inches per year and pastoral modes of subsistence are primary. The Maasai pastoralists of the Amboseli reserve inhabit a less arid zone and have access to somewhat greater resources than do the Turkana. Among the Karimojong (including the Dodos), rainfall, although erratic, averages about 20 inches per year. These conditions allow for a greater reliance on cultivation, a practice which substantially modifies human and livestock movement patterns.

If human populations and associated domestic herbivores (cattle, sheep, goats, and donkeys) are superimposed on the large variety and numbers of wild ungulates, then it is clear that tropical savanna ecosystems have a relatively high biological productivity. When wild ungulates exist apart from humans and domestic stock, wild ungulate productivity or biomass is high at between 12,000-17,000 kg/sq km (weight of animals per land area). Total biomass drops when domestic animals are added to the ecosystem to about 6,000 kg/sq km with domestic animals making up about 42 percent of this biomass.[14] Western and Dunne give a total large-mammal biomass value of 8,000 kg/sq km for the Amboseli game reserve, where Maasai and their domestic stock constitute about 50 percent of this value.[15] Thus a balance appears to be established between the human and domestic animals and the wild ungulates in the degree of exploitation of the savanna. In some ways this balance is struck by competition for available resources; in other ways the domestic and wild components of the ecosystem interact to facilitate exploitation.

East African Pastoralists Today

Among the pastoral peoples of East Africa, cattle are the most prized of pos-

13. R. Dyson-Hudson and N. Dyson-Hudson, "Subsistence Herding in Uganda," *Scientific American* 220, no. 2 (1969):76-89.

14. L. M. Talbot, "The Biological Productivity of the Tropical Savanna Ecosystem," in *The Ecology of Man in the Tropical Environment*, publication no. 4, n.s. (Morges, Switzerland: International Union for the Conservation of Nature and Natural Resources [IUCN], 1964), pp. 88-97.

15. D. Western and T. Dunne, "Location of Maasai Settlements in Relation to Geomorphology, Hydrology, and Vegetation in Amboseli Game Reserve, Kenya," *Human Ecology* (in press).

sessions. Often referred to as the "cattle complex" peoples, members of savanna pastoral tribes revere their herds. Cattle are intricately tied into the social organization, marriage practices, religion, and inheritance patterns; they are the primary indicators of wealth and prestige.[16] In addition to the sociocultural functions of cattle, livestock serve as the principal basis of subsistence in an ecosystem that is ideal for grazing and browsing of domestic and wild herbivores. Cattle, sheep, and goats convert the grasses and shrubs of the rangelands, which cannot be utilized directly by humans, into usable food products, such as meat, milk, and blood. At the same time, these food products are stored "on the hoof" and can be drawn upon when the need arises.

Perhaps one of the most curious and ingenious techniques of tapping this food reservoir without killing off livestock is the practice of bleeding animals at carefully timed intervals. Nearly all savanna peoples engage in this practice, and all varieties of livestock are bled. Cattle are usually bled by piercing the jugular vein in the neck with an arrow, after having tied a thong around the neck to distend the vein. Since cattle are large animals, between two and four quarts of blood (1.9-3.8 liters) can be drawn without seriously imparing the health of the animal. However, this can be done only two or three times each year for any given animal. Other techniques may be used for smaller animals, such as sheep, goats, and donkeys. Not all cattle, however, are bled regularly, except in the case of dire emergency. For example, milk-producing cows or young, growing calves will not be bled. If lactating cows were bled, milk production would drop, and the drawing of blood from immature animals seriously slows down the growth of the calves.

Many East African Pastoralists, such as the Karimojong who live in less arid zones, practice cultivation of crops. This requires a more sedentary pattern of existence with permanent or semipermanent settlements established close to lands suitable for cultivation. Plants commonly grown include sorghum, millet, corn, beans, peanuts, and pumpkins. In any given settlement women appear to perform much of the agricultural labor, although men will assist in many of the tasks.[17] Figure 5.7 illustrates the annual

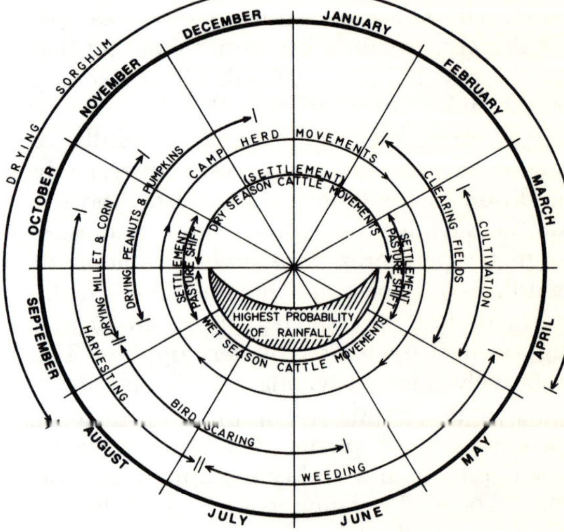

FIGURE 5.7. Annual cycle of subsistence activities of the Karimojong. (Data from Dyson-Hudson, R. 1972. "Pastoralism: Self Image and Behavioral Reality." In *Perspectives on Nomadism,* ed. W. Irons and N. Dyson-Hudson. Leiden: E.J. Brill, pp. 30-47.)

16. N. Dyson-Hudson, "Factors Inhibiting Change in an African Pastoral Society: The Karimojong of Northeast Uganda," *Transactions of the N.Y. Academy of Sciences* 24 (1962):771-801.

17. R. Dyson-Hudson, "Pastoralism: Self Image and Behavioral Reality," in *Perspectives on Nomadism,* eds. W. Irons and N. Dyson-Hudson (Leiden: E. J. Brill, 1972), pp. 30-47.

cycle of a group of Karimojong tribesmen. Field clearing and cultivation are performed beginning in February and March before the onset of the rainy season in April. The crop harvest begins in the latter months of the rainy season, while the drying of agricultural products for storage takes place during the most arid period of the year. Herds may be grazed close to the settlement or at some distance in camps that move in search of good pasturage and water, especially during the dry season. Men and boys tend the camp herds, which include the bulk of the cattle. Women, children, and aged adults remain in the settlement along with the majority of lactating cows that provide milk for children and adults alike.

Cattle movements depend on a complex series of strategy decisions that must be made by the camp herdsmen. Water supply, usually abundant during the rainy season (despite its unpredictability), is of primary importance during the long dry season. The areas where water frequently is available, however, may be dangerous because they are inhabited by hostile neighboring tribes or because they are zones where cattle diseases are prevalent (tick-borne east coast fever and tsetse-fly-borne trypanosomiasis in cattle). Herdsmen must also seek out rangelands that provide adequate forage for livestock; that is, the vegetation must have sufficient bulk and must provide an adequate protein:crude fiber ratio. Therefore calculated compromises must be struck continually during the dry season quest for watering places and forage. In many cases decisions of this nature are a matter of survival, not only for the camp herdsmen but also for those who remain in the settlement. Even in the best of dry seasons, cattle will show a marked body weight loss which produces a decline in milk production and limits the amount of blood that can be drawn for food.

Knowledge of the savanna ecosystem is quite sophisticated among pastoralists, and a remarkable degree of "environmental perception" is displayed. Among the Amboseli Maasai, the location of a settlement is a decision based on a number of important variables.[18] For example, settlements are generally found near ridges of hills that tend to have gradual slopes and minimum distance to the base of the hill. This appears to be a logical compromise, in that soil drainage is poor further downslope, and the area there becomes very muddy following heavy rains. Also, animal dung that accumulates further down the hillside does not decompose rapidly when saturated with water and may contribute to disease transmission in the cattle. An additional factor that encourages upslope location of settlements is based on soil color and temperature sensitivity in cattle. Light-colored soils, which tend to reflect and not trap heat, are found further downslope and appear to be associated with colder nighttime temperatures. Maasai claim that cattle are cold-sensitive, that cold conditions limit milk production. The principal reason why the ridge tops are not chosen for settlement sites appears to be based on cattle energetics. Cattle climbing up the longest slopes to settlements may expend three times more energy than required to climb the shorter slopes. Conservation of energy by cattle is especially crucial at the end of the dry season when body weight is very low, and cattle are all approaching their limits of survival.

Savanna pastoralists thus must exercise careful judgment in assessing seasonal variation in rainfall, forage availability and content, disease prevalence, and travel conditions. Furthermore, strategy decisions dealing with livestock maintenance and exploitation must be based on a great deal of

18. Western and Dunne, "Location of Maasai Settlements."

knowledge about the behavior and biology of the animals and about the savanna ecosystem. Food production must be maximized while the hazards associated with unpredictable rainfall and other factors must be minimized to insure survival of livestock and human populations.

Energetics of Savanna Pastoralists

East African Pastoralists exist on a tight energy budget. The combination of low rainfall, a long dry season, poor predictability of rain fall in any given year, and periodic droughts are severe limiting factors to productivity in the savanna. Because of these limiting factors, the human and livestock carrying capacity of the ecosystem is low in any given area, yet can be maximized by continual movements of settlements and herds in quest of water and good pasture. It appears quite clear that a sedentary or settled pattern of living among savanna pastoralists is incompatable with survival except in the wettest zones of the savanna. And it is in the zones with relatively high rainfall that intensive agriculture is practiced and animal husbandry is de-emphasized.

In zones where rainfall is adequate for some cultivation, such as the lands of the Karimojong and the Dodos, the economy is mixed with agriculture and herding playing equally important roles in providing food. A simplified energy-flow diagram of a mixed subsistence base practiced by the Karimojong is illustrated in figure 5.8. A number of important variables are left out of the diagram to reduce its complexity. For example, energy input from gathering of wild foods and occasional hunting is not included, nor is trade of livestock products or excess crops that may increase the system's overall energy input. Likewise, predation by large carnivores is a small but significant drain on cattle, sheep, and goats, and this should be considered in any comprehensive

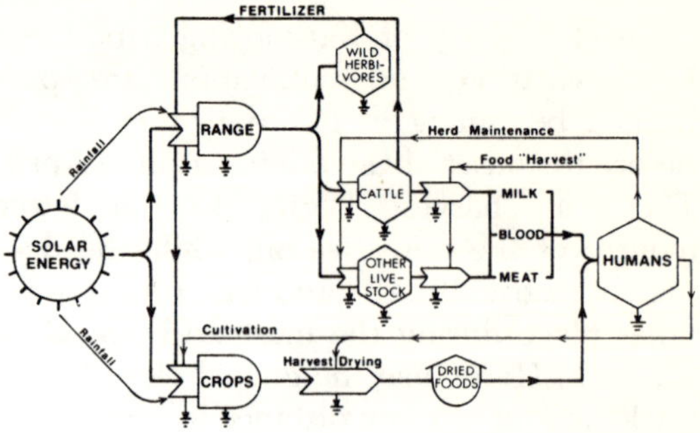

FIGURE 5.8. Energy flow in a Karimojong settlement.

model of the energetics of a pastoral population.

Unfortunately, the data necessary to quantify the energetics of any East African Pastoralist group are not yet available. Odum attempted to produce a flow diagram from the data of Deshler on the Dodos and concluded that only about one percent of the total food consumed was derived from milk, blood, and meat.[19] Such a low value is very unlikely even in years where agricultural productivity is maximal. Based on unpublished data gathered by N. Dyson-Hudson on the neighboring Karimojong, estimates of livestock food-energy productivity suggest that herds contribute substantially more food than the 1 percent from Odum's calculations. These estimates are given in table 5.1 for the Karimojong. The average individual daily food requirements (1,800 kcal) are based on any given settlement having a large proportion of children, and the body weight of individuals at all ages being lower than that of Europeans. Milk and blood yields are based on actual data gathered from the Karimojong; meat yield, on a single adult animal

19. W. W. Deshler, "Native Cattle Keeping in Eastern Africa," in *Man, Culture, and Animals*, eds. A. Leeds and A. P. Vayda, publication no. 78 (Washington, D.C.: American Association for the Advancement of Science, 1965), pp. 153-68.

TABLE 5.1

ESTIMATED FOOD ENERGY UTILIZED BY MEMBERS OF A KARIMOJONG SETTLEMENT

Settlement population size		25 (men, women, and children)
Human food requirements		1,800 kcal/person/day (all ages) 45,000 kcal/total settlement/day
Herd size (cattle only)	100	
Ratio (cattle:person)	4:1	
Herd structure and exploitation		
24 milch (lactating) cows		— milked, not bled
35 dry (nonlactating) cows		— bled
18 adult males (3 bulls, 15 steers)		— steers bled
23 calves		— not bled
	Energy Yields	
Milk		— 1 liter/day/milch cow × 300 days × 24 cows = 7,200 liters/year × 680 kcal/liter of milk = 4,896,000 kcal/year
Blood		— 400 liters/year × 500 kcal/liter of blood = 200,000 kcal/year
Meat		— 150 kilograms/year × 3,000 kcal/kg of meat = 450,000 kcal/year
Total kcal		— 4,896,000 (milk) + 200,000 (blood) + 450,000 (meat) = 5,546,000 kcal food/year, or 15,195 kcal/day available to members of the settlement

(Unpublished data from Neville Dyson-Hudson.)

dying or being slaughtered in a given year. The food energy of 15,195 kcal/day derived from cattle thus provides about 34 percent of the total estimated requirements of the members of the settlement.

Since most Karimojong pastoralists keep goats, sheep, and donkeys in addition to cattle, the proportion of food from animal products is probably substantially greater than the 34 percent noted above. All animals can be bled should the need arise, and female goats and sheep can be milked as well. The Karimojong and Dodos, who practice a mixed economy of cultivation and herding, probably do not exceed a diet comprising 50 percent animal products in light of the known ratios of livestock to people (about 4:1 for cattle and 3:1 for goats and sheep). For a savanna population that relies more heavily on its animals for food, the ratios must be higher. This holds for Kenya Maasai who show ratios of about 6:1 for cattle, and 5:1 for goats and sheep. However, among the Turkana who inhabit some of the most arid regions of Kenya, cattle ratios are low. Apparently, Turkana compensate by incorporating larger numbers of goats as well as camels; both animals, of course, are more tolerant of arid conditions. In addition, reported numbers of cattle owned by the Turkana are probably underenumerated, since camp herds tend to be widely dispersed because of the scarce resources.

A note of caution should be inserted here regarding the distribution of food by season and by age and sex of members of any given settlement. For example, among the Karimojong during the dry season, young men tending the camp herds (at some distance from the home settlement) may subsist wholly on animal products, while the children and older adults at the settlement will rely on only plant foods and some milk. During the wet season, more animal foods will be available to those who remain at the settlement, and camp herdsmen will have greater access to cultivated plant foods. Hence at any given time males are likely to be consuming more protein-rich animal foods than are females.

Health and Biology of Savanna Pastoralists

Research on the human biology of savanna pastoralists is scanty. However, studies of populations from other parts of Africa can be used in conjunction with the work that has been done to predict patterns of growth, disease, temperature regulation, and nutritional relationships among savanna dwellers.

There are a variety of interesting relationships between human growth of pastoral nomads and the savanna ecosystem that are still to be explored. Roberts studied child growth in a number of African populations including the pastoral Dinka (a Nilotic people from the southern Sudan) and compared this data with information gathered by others on London and pastoral Maasai children.[20] Dinka and Maasai children between the ages of four and sixteen years are quite comparable to the British children in height but considerably lighter in body weight at all ages, including adulthood. This linearity of physique, apparently unrelated to nutrition and a common feature of savanna pastoralists, may be related, as Roberts suggests, to the hot-dry climate of this environmental zone. A linear or thin physique can be adaptive under hot-dry conditions by maximizing the body surface area:body weight ratio, which facilitates loss of excess heat that is stored by the body. Hiernaux, from growth studies of pastoral Rwanda Tutsi, agrees with Roberts's suggestion.[21] Although nutrition does not appear to be a major factor in the linear physique of savanna pastoralists, nutritional and disease variations may influence growth patterns on a seasonal basis. In East Africa, it is well known that young cattle grow rapidly (as measured by weight gains) during the wet season when forage is abundant and rich in protein, but lose weight during the dry season when good forage and water resources are scarce. Thus since cattle are a primary source of food for growing children, it is likely that children may display similar seasonal growth variations associated with food availability. Wet-dry seasonality in disease incidence could also influence human growth. An excellent example of seasonal influences on growth in West African children from Gambia is illustrated in figure 5.9. Infections in this area are much more common during the rainy season and contribute to a decline of growth in body weight, sometimes even producing a loss of weight in young children. It is difficult to predict what influence seasonality would have on children of pastoral nomads, but some seasonal effects probably would be felt.

One of the most serious problems facing East African Pastoralists is the presence of tsetse-fly borne trypanosomiasis, a disease that affects both cattle and people.[22] Lands

20. D. F. Roberts, "Effects of Race and Climate on Human Growth as Exemplified by Studies on African Children," in *Human Growth,* ed. J. M. Tanner (London: Pergamon Press, 1960), pp. 59-72.

21. J. Hiernaux, "Height/Weight Relationship During Growth in Africans and Europeans," *Human Biology* 36 (1964):273-93.

22. J. Ford, "Interactions Between Human Societies and Various Trypanosome-Tsetse-Wild Fauna Complexes," in *Human Ecology in the Tropics,* eds. J. P Garlick and R. W. J. Keay (London: Pergamon Press, 1970), pp. 81-97.

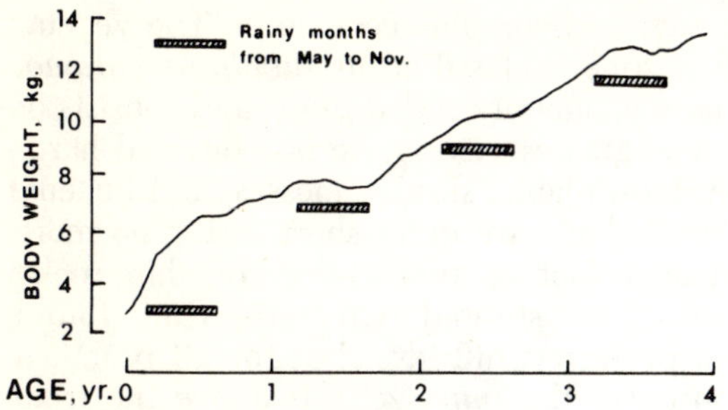

FIGURE 5.9. Seasonal variation in growth of children from a village in Gambia. (Adapted from Thomson, A.M. 1970. "The Development of Young Children in a West African Village: A Study in Human Ecology." In *Human Ecology in the Tropics*, ed. J.P. Garlick and R.W.J. Keay. Oxford: Pergamon Press, p. 103.)

infested with tsetse-flies are rich in livestock forage since the flies require a moderate amount of vegetation cover for survival. Hence any exploitation of lands where the tsetse-fly is present carries a risk of infection in cattle and humans. Distribution of the tsetse-fly and other diseases (east coast fever) severely limits grazing movements, although in cases of prolonged drought, such areas will be exploited out of necessity since the more serious risk of starvation outweighs the risk of disease. These interactions are further complicated by gradual shifts in the distribution of tsetse-flies and disease outbreaks from time to time.

Body-temperature regulation and maintenance of water and salt balance is a problem among savanna pastoralists, especially those who reside in the hotter and more arid zones. Wyndham investigated heat stress responses in South African Bushmen and Bantu and Algerian Arabs from the Sahara, but no studies have been conducted to date on savanna nomads from the equatorial zone.[23] He found that Bushmen and Bantu were better acclimatized to heat (than Europeans) as reflected by their lower rectal temperatures and lower heart rates when exposed to hot conditions. Similar if not greater levels of heat acclimatization would be expected among savanna nomads.

Relationships among nutrition, health, and human growth are difficult to establish for savanna pastoralists because of the paucity of data on these peoples. Surveys were done in Kenya and Tanzania, but the focus was on Bantu agricultural peoples and not herders.[24] Among Bantu agriculturists, child growth was found to be retarded in many cases, generally attributable to low protein intake. Savanna pastoralists should be interesting groups by comparison, since protein intakes from milk, blood, and meat would be expected to be considerably greater than those of agriculturists due to the pastoralists' heavy dependence on livestock.

PERUVIAN ALTIPLANO INDIANS

High-altitude populations at elevations above 3,000 meters are exposed to a wide spectrum of stresses associated with climate, vegetation, isolation, and other factors. Many of these stresses cannot be reduced even with sophisticated Western technology. Of the major high altitude populations inhabiting the Himalayan and Andean Plateaus, only the latter has been studied in any depth. The material that follows will focus on the IBP work conducted in the community of Nuñoa, Peru (see figure 5.10).

23. C. H. Wyndham, "Southern African Ethnic Adaptation to Temperature and Exercise," in *The Biology of Human Adaptability*, eds. P. T. Baker and J. S. Weiner (Oxford: Clarendon Press, 1967), pp. 201-44.

24. H. Kraut and H. D. Cremer, eds., *Investigations into Health and Nutrition in East Africa* (Munich: Weltforum Verlag, 1969).

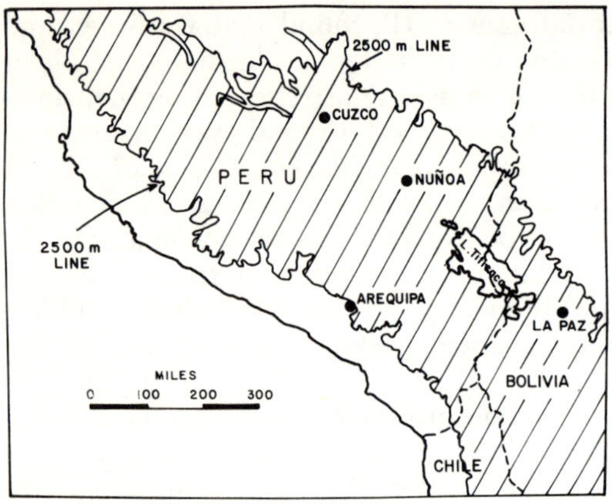

FIGURE 5.10. Map of the Peruvian altiplano.

The Andean Altiplano Ecosystem

The Andean altiplano of Peru and Bolivia is a high plateau surrounding Lake Titicaca. The relief of the altiplano is not as severe as that found in the northern and central Peruvian Andes, but rather it is characterized by broad valleys dissected by meandering streams and flanked by mountains and hills that rise as much as 1,000 meters above the valley floors. Few of the valleys of this plateau are below the elevation of 3,700 meters (12,000 feet) above sea level; habitable valleys can be found up to 5,000 meters (16,400 feet), and some mountain peaks rise to more than 6,100 meters (20,000 feet). The climate of this high Andean zone is truly as formidable as the topography. At these elevations, the air is rarefied and contains substantially less oxygen than at sea level; air temperatures are frequently below freezing; and because of the rarefied air, solar radiation is intense. Being within the Tropic of Capricorn latitudinal zone, the permanent snow line is at about 5,200 meters (17,000 feet).

Despite what might be considered a harsh environment, the altiplano supports a rich and complex ecosystem. The vegetation zone is classified as *subalpine paramo*, more commonly called *puna*, and consists of bunchgrasses (*ichu*), herbaceous and hard-cushion plants, shrubs, mosses, and lichens. Most plants are quite short, rising no more than a foot or two above the thin rocky soil. Some stunted native trees are found on protected hillsides, but for all practical purposes the *puna* zone is above the tree-line and is basically a natural alpine grassland. Wild animals include varieties of small mammals, some deer, and a few small herds of free-ranging Andean camelids (*vicuña* and *guanaco*). Hundreds of varieties of birds can be found throughout the altiplano.

At high elevations in southern Peru the climate is uniform throughout the year because of the proximity to the equator. During the daylight hours, air temperatures are moderate, while at night, temperatures drop rapidly, often to levels well below freezing. As a result, the daily (diurnal) temperature variation is great. Seasonal climatic variation is largely the result of the monsoon pattern of wet and dry periods. In the altiplano the rainy season extends from about September to April, the dry season from May to August. The highest and lowest temperatures of the year are usually recorded during the dry season when cloud cover is light. Solar radiation is high and temperatures are warm during the day, while at night, temperatures fall because the lack of clouds facilitates the loss of heat radiation to the sky. During the rainy season, heavy cloud cover moderates temperatures, producing cooler days and less cold nights. Figures 5.11A and 5.11B illustrate temperature and rainfall patterns for the town of Nuñoa at 4,000 meters (13,150 feet) in southern Peru. These climatic patterns are independent of altitude, yet average yearly temperature drops about 1°C for each 100 meters increase in elevation (see figure 5.11C).

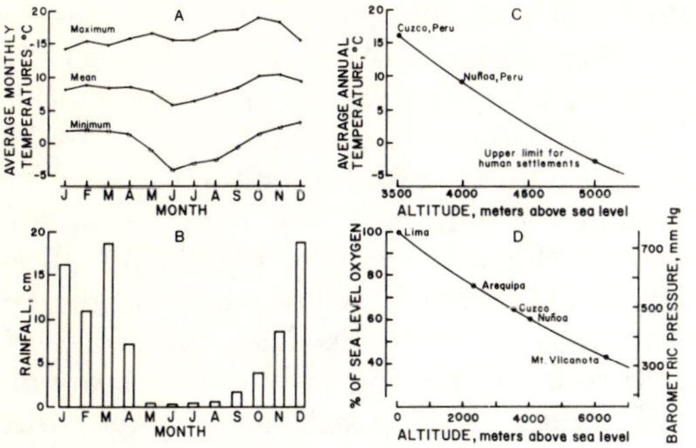

FIGURE 5.11. (A and B) Weather data from the town of Nunoa (1968), (C) relationship between altitude and temperature and (D) relationship between altitude and percentage of sea level oxygen.

Perhaps more important because of its influence on all living forms is the reduced oxygen pressure found at high altitudes. As air temperatures decline with increasing altitude, so do air pressures (see figure 5.11D). For example, air at 4,000 meters above sea level contains about 60 percent of the oxygen in air at sea level. The combined effect of low temperatures and air pressure tends to limit growth of natural vegetation and cultivated plants, and at the same time has a profound effect on animal and human populations continually exposed to these environmental stresses. Cold and the hypoxia produced by low oxygen pressure are two of the most important limiting factors in the dynamics of the altiplano ecosystem.[25]

Altiplano Indians Today

An estimated 500,000 subsistence farmers and herders are equally divided within the altiplano regions of Peru and Bolivia. The remaining sections of this case study will look at the Peruvian altiplano and more specifically at studies conducted in Nuñoa, a political district with about 8,000 persons located about 125 kilometers northwest of Lake Titicaca (see figure 5.10).

In the altiplano of Peru there tend to be three roughly defined social classes: the *mestizo,* the *cholo,* and the *indigena* (Indian). Members of the *mestizo* (upper class) wear western clothing, speak Spanish as the primary language, and have a level of education beyond elementary school. Most shopkeepers, hacienda (farm) owners, school teachers, merchants, and political officials are classified as *mestizos.* The *cholo* is an upwardly mobile Indian who has acquired the use of Spanish, and who wears a distinctive style of manufactured clothing. *Cholos* are generally small-shop owners, craft specialists, and hacienda foremen. The *indigena* class makes up the bulk of the population of the altiplano. These are the farmer-pastoralists who can be identified by their use of Quechua as a primary language, by their use of *coca* (a mildly narcotic leaf that is chewed), and by their wearing of homespun wool clothing. Assignment to one of the three social classes does not necessarily relate to the biological origins of a group. However, members of the *cholo* and *indigena* social classes are usually native Andean Indians biologically with little European admixture.

For the majority of the rural Indians, the hacienda system is the dominant way of life. Under this system of privately owned farms and estates, the Indian is a landless tenant who must perform services for the hacienda in exchange for the privilege of living on

25. For a comprehensive description of the altiplano ecosystem, see, R. B. Thomas and B. Winterhalder, "The Physical and Biotic Environment of Southern Highland Peru," in *Man in the Andes: A Multidisciplinary Study of High Altitude Quechua,* eds. P. T. Baker and M. A. Little (Stroudsburg, Pa.: Dowden, Hutchinson, and Ross, 1976).

the lands.[26] The economy of the hacienda and its Indian residents is centered around herding llamas, alpacas, and sheep. These animals contribute hides and skins for bedding, wool for clothing and blankets, dung for fuel and fertilizer, and meat products for food. The hides and animals are often sold through the market systems, although the greater part of the family income is derived from the sale of wool in December. With this income, dyes for yarn, grain foods (wheat, barley), *coca,* alcohol (for drinking), and a few small luxuries can be purchased. Figure 5.12 represents some of the exchanges that take place. All of the meat and the wool for basic clothing are provided for by the herds.

Every Indian family owns some llamas, alpacas, and sheep, and many own at least one horse. Llamas and horses are used to transport goods and people respectively, while alpacas and sheep are the most important sources of meat and wool. Each day the herds are moved out from the family settlement in the early morning and returned before dark in the evening. All members of the family participate in pastoral activities from time to time, and young children are often responsible for large flocks of animals. During the dry months of May through August, herds may be moved to higher and wetter pastures.

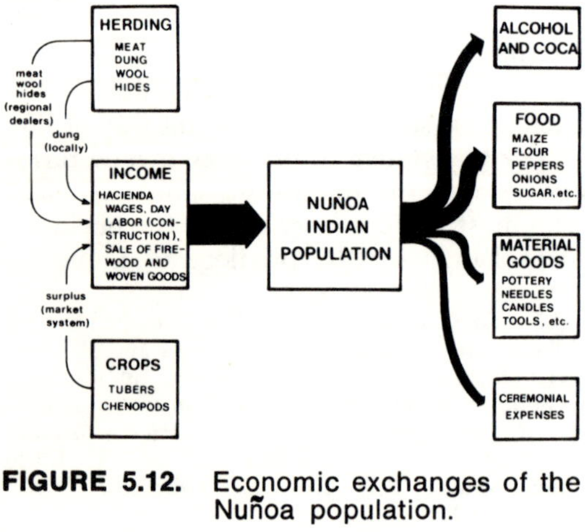

FIGURE 5.12. Economic exchanges of the Nuñoa population.

Lands below about 4,300 meters (14,000 feet) are utilized for agriculture. Above this elevation nighttime frosts, common at any time of the year, prevent plant cultivation. Hence at elevations below this level the Indian engages in a mixed economy of agriculture and herding, while the economy shifts fully to pastoralism at higher altitudes. Several weeks of intensive work is devoted each year to agricultural activities, with about equal effort expended in planting and harvesting. Frost resistant crops grown in the altiplano include white potatoes (the staple), other tubers (*año, oca*), and the native grains *quinoa and cañihua* (chenopods). Planting occurs in September and October at the onset of the rainy season, while crops are harvested in March, April, and May. All agricultural work is carried out with hand and foot tools. A diagram illustrating subsistence activities throughout the year is given in figure 5.13.

Although many Indians living above the maximum altitude of cultivation grow none of their own food, vertical trading is commonplace, and plant foods are an important part of all diets. Dried potatoes (*chuño negro* and *moraya*), fresh potatoes, *quinoa, cañihua,* corn, barley, and flour are included in the usual dietary regimen. Animal products include *charqui* (dried salted meat), fresh meat, rendered fat (*cebo*), and occasionally milk, cheese, and eggs. A typical cooked meal consists of a thick porridge containing dried potatoes, flour or ground barley, onions, some spices, and either meat or *cebo.* The variations on this standard theme depend on the foods available at different times of the year, the relative prosperity of the family, and the alti-

26. This system is presently being changed in Peru through governmental programs of agrarian reform.

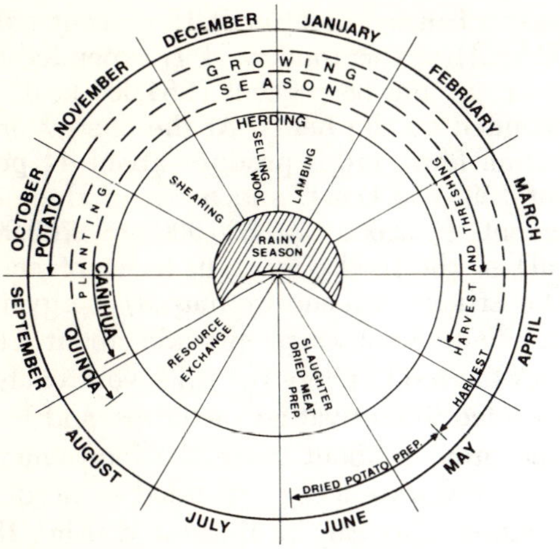

FIGURE 5.13. Annual cycle of Nuñoa Indians. (Data from Thomas, R.B. 1973. "Human Adaptation to a High Andean Energy Flow System." *Occasional Papers in Anthropology,* no. 7. University Park, Pa.: Pennsylvania State University.)

tude of residence. For example, dietary consumption of the chenopod grains increases following the harvests in March and April, as does the consumption of potatoes in April.

The protection against cold afforded by clothing and shelter would be considered poor by Western standards. A rural family settlement usually consists of three or more one-room huts for sleeping, cooking, and storage respectively. The huts are grouped around a central dirt patio with dry stone-walled corrals for the animals adjacent to the settlement. Each small hut is constructed of field stones and mud mortar with a roof of bunched grass. Only in the kitchen hut is any warmth provided, and this results from the cooking fires that are kindled in the early morning and early evening. The huts provide little insulation, so that during the coldest hours at night, the temperature difference between inside the sleeping hut and outdoors is a matter of only 1°-2°C.

Virtually all clothing worn by the Indian pastoralist is homespun or knitted from wool. Men and boys wear one or two pairs of trousers, a shirt, a vest, a sweater, and a short jacket. A knitted cap and heavy poncho are reserved for the coldest times of the day or night. Women and girls wear several skirts, a blouse, a short jacket or vest, and a shawl. Men frequently wear rubber-tire sandals, while women and children more commonly go barefoot. No gloves, stockings, or any other protection for the hands and feet are worn.

Energetics of the Nuñoa Population

In Nuñoa, there are three major trophic levels within the so-called human-managed component of the ecosystem. The first level includes the *puna* wild native grasses and the cultivated crops that make up the greater part of the *producers.* The second level comprises the domestic herbivores—the *primary consumers*—including the camelids (alpacas and llamas), sheep, horses, and cattle. At the third level of *higher-order consumers* is the human population. Other higher-order consumers, such as wild carnivores, play a relatively minor role in the "managed" component of the ecosystem. The Indian subsistence economy, although on a tight energy budget, is an efficient one in the sense that a reasonable degree of balance can be maintained within the ecosystem. For example, most of the Indian's clothing, food, fertilizer, and fuel needs can be provided locally,

Beginning at the producer trophic level, *puna* grasses, and occasionally domestic grains, provide forage for the livestock. The livestock in turn provide meat, milk, dung, wool, hides, and transport for the native population. Some of these products, particularly meat, wool, and hides, are traded through regional dealers to provide income for the purchase of food or other commodities (see figure 5.12). Livestock dung is an important by-product of herding. It provides high-energy fuel for cooking and rich fertilizer either for *puna* grass production or for use in agriculture.[27] Finally, frost-resistant and altitude-tolerant cultivated grains and tubers provide some of the staples of the native diet.

R. B. Thomas has gathered substantial information on energy flow through the Nuñoa native population and has constructed a model that illustrates some of the comments made above.[28] The model is shown in figure 5.14. Here the relationships among (1) *energy production* (the value of animal and plant foods), (2) *energy consumption* (the value of food consumed), and (3) *energy expenditure* (the amount of work put into food production) are standardized in terms of their respective caloric values (in kcal) for an Indian family of six persons.

Dealing first with energy expenditures—the work involved in cultivation and herding—we can see that about 12 percent (51,800 kcal) of this total work is expended on the crops, and 88 percent (384,000 kcal) is expended on the herds. At the energy production level, crops produce about 44 percent (595,000 kcal), and herds produce 56 percent (222,000 + 538,000 = 760,000 kcal) of the food. Hence in terms of energetic efficiency (energy output/energy input), cultivated crops provide about 11.5 times the food energy per unit work energy expended in cultivating the crops, and livestock provide about twice the food energy per unit work energy expended in herding activities. This can be illustrated using the actual caloric values given on page 67.

27. B. Winterhalder, R. Larsen, and R. B. Thomas, "Dung as an Essential Resource in a Highland Peruvian Community," *Human Ecology* 2 (1974):89-104.

28. R. B. Thomas, "Energy Flow at High Altitude," in *Man in the Andes: A Multidisplinary Study of High Altitude Quechua*, eds. P. T. Baker and M. A. Little (Stroudsburg, Pa.: Dowden, Hutchinson, and Ross, 1976).

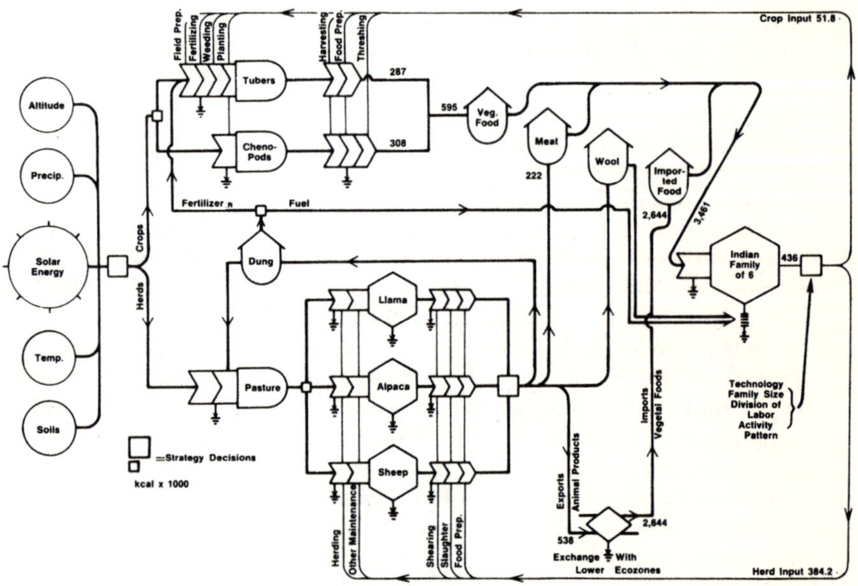

FIGURE 5.14. Energy flow in a Nuñoa Indian family. (Adapted from Thomas, R.B. 1974. "Human Adaptation to Energy Flow in the High Andes: Some Conceptual and Methodological Considerations." In *Energy Flow in Human Communities*, eds. P.L. Jamison and S.M. Friedman. University Park, Pa.: Pennsylvania State University.)

$$\text{Energetic efficiency of cultivation} = \frac{\text{food produced}}{\text{agricultural work}} = \frac{595{,}000 \text{ kcal}}{51{,}800 \text{ kcal}} = 11.5$$

$$\text{Gross energetic efficiency of herding} = \frac{\text{food produced + exports}}{\text{herding work}} = \frac{222{,}000 + 538{,}000 \text{ kcal}}{384{,}200 \text{ kcal}} = 2.0$$

These calculations, however, only give a gross picture of the energetic efficiency of the human subsistence system. By exchanging animal products through trade for food products with a higher caloric value, the overall energetic efficiency of herding can be markedly increased.

$$\text{Net energetic efficiency of herding} = \frac{\text{food produced + imports}}{\text{herding work}} = \frac{222{,}000 + 2{,}664{,}000 \text{ kcal}}{384{,}200 \text{ kcal}} = 7.5$$

This shifts the percentage of food produced by cultivation from 44 percent to 17 percent, and the percentage of food produced by herding from 56 percent to 83 percent. Moreover, if the energy value of the dung produced by the livestock and required for cultivation were incorporated into the model, the energetic efficiency of herding would increase even more.

These relationships demonstrate the utility of a mixed economy in the Andean altiplano. Herding of livestock provides a stable resource, invaluable dietary protein from animal foods, fertilizer for agriculture, and an important trade commodity. Agricultural productivity on the other hand is insufficient alone to provide all of the food needed for survival, can be limited due to the dangers of crop failure, and in turn dependent on animal products.

Health and Biology of Altiplano Indians

The health of Altiplano Indians can be considered good in some ways and poor in others. For example, as in many peasant populations, health care is not readily accessible to most members of the population, and infant and early childhood mortality is high. However, the Indians tested in Nuñoa are physically fit and are able to perform exercise tasks at least equally as well as sea-level natives. Dental health is excellent, and nutritional status is adequate when the diet consists largely of native foods. When high-carbohydrate Western foods replace native foods in the diet, imbalances and deficiencies commonly occur.[29] There is also a great deal of seasonal variation in diet that results from the annual patterns of food production. Fresh potatoes, for example, are an important part of the diet from June through October, while dried potatoes are increasingly consumed during the remainder of the year (see figure 5.13). Although the native diet appears adequate for adults leading a normally active life, there are indications that protein and calorie intake may be insufficient during childhood growth and for women who are either pregnant or are nursing infants.

Respiratory diseases, such as pneumonia, influenza, and tuberculosis, are quite common in high-altitude populations. In Nuñoa, nearly 60 percent of all deaths, especially among infants and children, are caused by respiratory diseases. This high incidence of respiratory ailments may be related somehow to altitude-induced hypoxia, since gastrointestinal diseases and infant diarrhea are major causes of illness and death in most non-Western peasant societies.

29. E. Picon-Reategui, "Nutrition," in *Man in the Andes: A Multidisplinary Study of High Altitude Quechua,* eds. P. T. Baker and M. A. Little (Stroudsburg, Pa.: Dowden, Hutchinson, and Ross, 1976).

Heart disease among adults is relatively uncommon, partly resulting from the short life expectancy of the population, but probably in part because of the virtual absence of high blood pressure, or hypertension. In the United States, hypertension is common among both males and females beyond the ages of forty-five or fifty years. By contrast, Nuñoa natives in their sixties and seventies have blood pressures that are well within the normal range for a 20-year-old in the United States. Although high altitudes may be a factor in the absence of hypertension among Andean Indians, it is clear that physical fitness and the low-fat diet of natives reduce the risk of developing this disease.

The growth of Nuñoa Indians is relatively slow; full adult size is not achieved until between twenty and twenty-five years of age.[30] In addition, sexual maturation is late by U.S. standards, and the period of rapid adolescent growth charcteristic of most other populations does not occur. For human growth to be successful in the altiplano environment, the young at all ages must combat a variety of environmental stresses. Such stresses as hypoxia, disease, limited food, and cold can be both reduced and intensified by the cultural system from time to time. However, biological patterns of adaptation are equally important in reducing the effects of these stresses.

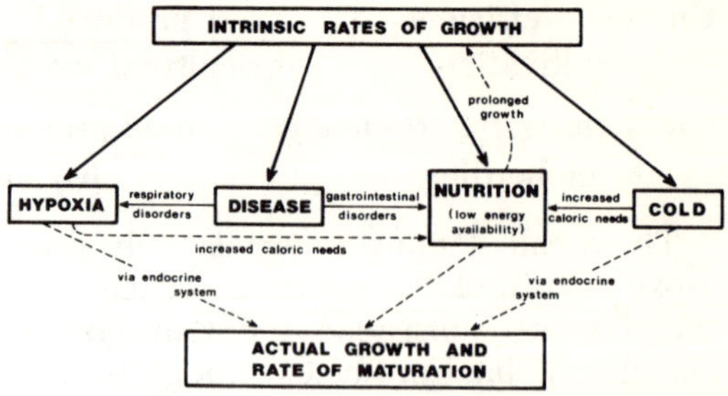

FIGURE 5.15. Environmental influences on growth of Nuñoa children.

Figure 5.15 illustrates how four stresses might influence genetic or intrinsic rates of growth. Homespun wool clothing and adequate bedding tend to reduce the effects of the cold temperatures found in the altiplano. On the other hand the limited fuel resources for heating and poorly insulated native huts permit a substantial degree of cold exposure, especially among children. Such cold exposure increases the individual's food-energy requirements and may actually worsen nutritional conditions during periods when food is limited. Correspondingly, stomach and intestinal infections will slow down, or limit, absorption of nutrients and even such mild respiratory diseases as the common cold can act to increase the stress associated with reduced atmospheric oxygen. Children are especially sensitive to stresses of all kinds, and growth can actually be halted for a period of time if the stress is large enough. Thomas suggests that the slow and prolonged growth of Altiplano Indians and their small adult body size are adaptations to the limited food resources and act to reduce the metabolic needs of individuals.[31] Disease, hypoxia, and cold temperatures may be additional factors contributing to this slow growth pattern.

Because of the demands of survival, the Indian must perform work to provide shelter, clothing, and food despite the presence of hypoxic and cold stresses. Increased activity or such work as plowing potato fields or walking long distances over rugged terrain increases one's oxygen needs. Supplying these needs in the oxygen-poor altiplano

30. A. R. Frisancho and P. T. Baker, "Altitude and Growth: A Study of the Patterns of Physical Growth of a High Altitude Peruvian Quechua Population," *American Journal of Physical Anthropology* 32 (1970):279-92.

31. R. B. Thomas, *Human Adaptation to a High Andean Energy Flow System,* Occasional Papers in Anthropology, no. 7 (University Park, Pa.: Department of Anthropology, Pennsylvania State University).

environment is a serious problem that appears to have been solved reasonably well by the high-altitude native. Physiologically and morphologically, Indians are able to take in oxygen and deliver it more efficiently to the tissues than sea-level residents. Much of this advantage seems to derive from a lifelong residence at high altitude, although a part of the advantage may have a genetic basis and hence be built into the biological structure of the population.[32] In addition to biological modes of adaptation to hypoxia, behavioral and cultural means are employed to reduce the additional stress of labor at high altitude. For example, children often carry out tasks that would require more work if performed by adults; a great deal of the daily subsistence activity is not strenuous.

As already noted, the degree of cold stress experienced by the native is influenced by the insulation of clothing and bedding and low availability of fuel. Cold stress may be increased by the requirements of subsistence activities that must be performed under a wide variety of weather conditions, including high wind chill and precipitation as rain, hail, or snow. From tests of cold tolerance conducted in the laboratory in Nuñoa, Indians can produce more body heat while at rest (basal metabolism) and maintain body temperatures at higher levels when exposed to cold than sea-level residents.[33] Moreover, Indians can maintain hand and foot temperatures at reasonably comfortable levels when exposed to cold, in marked contrast to the reactions of sea-level residents.[34] Only a few of these relationships have been touched on, while many still remain unexplored.

NEW GUINEA HIGHLANDERS

The large island of New Guinea offers unique opportunities for the study of human cultural and biological variation. With a total population of between two and three million, there are perhaps four hundred language groups and literally thousands of self-contained local populations. Those populations living in the central mountain range show somewhat less variability than in other regions of New Guinea, but are manageable within the ecological framework presented in this work. At the same time, it is important to take advantage of the degree of uniformity of this region and to discuss a limited number of attributes in detail.

The million or so people living in montane New Guinea are exposed to a wide variety of environmental problems. Some problems are peculiar to interior and high-altitude zones, including those associated with earth and atmospheric conditions, vegetation and fauna, disease, and competition with other humans. For example, some of the most critical behavioral and physiological responses of these human populations are associated with the limited availability of high-quality protein and other nutritional elements. Satisfying these needs has involved the expansion and intensification of resource-management strategies through the application of technology,

32. M. A. Little and P. T. Baker, "Environmental Adaptations and Perspectives," in *Man in the Andes: A Multidisplinary Study of High Altitude Quechua*, eds. P. T. Baker and M. A. Little (Stroudsburg, Pa.: Dowden, Hutchinson, and Ross, 1976).

33. P. T. Baker et al., "Temperature Regulation at High Altitude: Quechua Indians and U.S. Whites during Total Body Cold Exposure," *Human Biology* 39 (1967):155-69; R. B. Mazess et al., "Oxygen Intake and Body Temperature of Basal and Sleeping Andean Indians at High Altitude," *Aerospace Medicine* 40 (1969):6-9.

34. M. A. Little, et al., "Population Differences and Developmental Changes in Extremity Temperature Responses to Cold Among Andean Indians," *Human Biology* 43 (1971):70-91.

the redistribution of population, and the control of fertility.

A number of populations of the New Guinea highlands (of Papua, New Guinea) have been the subject of human ecological and biological research. In fact this area has played an important role in developments in a number of complimentary fields. These include rigorous ecological investigations of agricultural systems, studies of dietary intake and energy expenditure sponsored by the South Pacific Commission and the Human Adaptability section of the IBP, and research on disease epidemiology, including malaria and the latent virus of kuru.[35] This case study draws largely from the research on three tribal groups of the New Guinea highlands: the Miyanmin who reside in the northwestern part of the central mountain range in Papua, New Guinea; the Marings who inhabit the northern edge of the western highlands, and who have been the subject of a major interdisciplinary project initiated by A. P. Vayda; and the Enga from the central highlands who have been studied by the anthropologist Mervyn Meggitt and the geographer Eric Waddel. These groups and others to be cited may be located in figure 5.16.

Tropical Montane Ecosystems

The highland region of Papua, New Guinea is one of the most rugged populated areas of the world. The central cordillera consists of a four-hundred-mile complex of mountain ranges, high valleys, and plateaus. Terrain is characteristically steep, with sheer slopes, sharp ridges, landslip scars, and swift rivers both in narrow valleys or ravines and in the broad valley floors. Among the ranges of mountains included are (from west to east) the Stars, Hindenbergs, Victor Emmanuels, Thurnwalds, Shraders, Bismarcks, Kratkes, and Owen Stanleys. Broad intermontane valleys, which contain the bulk of highland population, include Ifitaman (Telefomin), Tari, Mendi, Lai-Enga, Wahgi-Chimbu, and Asaro, each with an elevation of about 1,500 meters (5,000 feet) above sea level. All of the ranges named exceed 3,000 meters (9,800 feet) in altitude. Drainage is provided for the northern slopes by tributaries of the Sepik and Ramu Rivers, and for the southern slopes by tributaries of the Fly, Strickland, Kikori, and Purari Rivers.

The Enga people, numbering some 140,000, dwell in a series of valleys to the west of Mount Hagen, with the Lai Valley the demographic center. They occupy the flatter terrain and terraces from 1,500 meters (5,000 feet) in the floors and the surrounding slopes up to 2,700 meters (8,800 feet). The Marings live in the narrow Simbai and Jimi Valleys in the Bismarck Range, exploiting most intensively the altitudinal range from 670 meters (2,200 feet) to 1,500 meters (5,000 feet). The Miyanmin live in the valleys of the Fak, August, and May Rivers (headwater tributaries of the Sepik) in the Thurnwald Range and the western

35. H. Brookfield and P. Brown, *Struggle for Land: Agriculture and Group Territories Among the Chimbu of the New Guinea Highlands* (Melbourne: Oxford University Press, 1963); Roy Rappaport, *Pigs for the Ancestors* (New Haven: Yale University Press, 1968); E. H. Hipsley and N. E. Kirk, *Studies of Dietary Intake and Expenditure of Energy by New Guineans,* technical paper no. 147 (Noumea: South Pacific Commission, 1965); G. A. Harrison and R. J. Walsh, eds., "A Discussion on Human Adaptability in a Tropical Ecosystem: An I.B.P. Human Biological Investigation or Two New Guinea Communities," *Philosophical Transactions of the Royal Society of London (Series B) Biological Sciences* 268, no. 893 (1974):221-400; R. Black, "Enlargement of the Liver Due to Malaria in Natives of the South-West Pacific Area," *Transactions of the Royal Society of Tropical Medicine and Hygiene* 48 (1954):226-33; D. C. Gajdusek and M. Alpers, "Genetic Studies in Relation to Kuru. I. Cultural, Historical, and Demographic Background," *American Journal of Human Genetics* 24, supplement (1972):1-38.

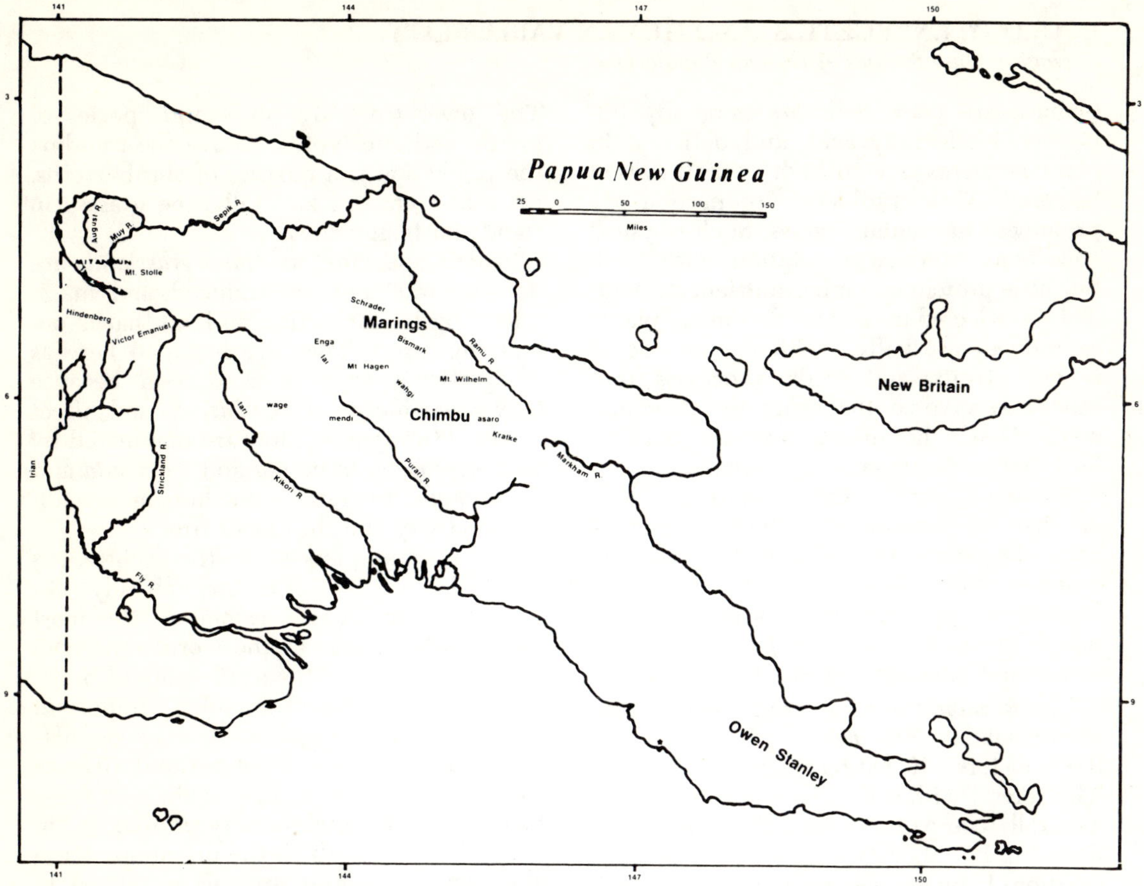

FIGURE 5.16. Map of the highlands of Papua, New Guinea.

slopes of Mount Stolle. In this area the central range does not rise steeply from the plains but instead ascends in a series of foothills. Hence although the range of more intensive exploitation by the Miyanmin varies from place to place, it can be roughly described as falling within the 450 meters (1,500 feet) to 1,400 meters (4,000 feet) zone.

Atmospheric conditions in the highlands of Papua, New Guinea show considerable variation. In combination, such factors as elevation and landform produce distinct local microclimates on a narrow scale. The resulting local conditions may include the occurrence of frosts, gully winds, temperature inversions, convectional rain, and rain "shadows." For the present purposes a few generalities can be made. The highland climate is seasonal; the biotic community (including man) displays seasonal responses. Rainfall in the highlands is high at 100-200 inches (254-508 cm) per year and is patterned according to three seasons: the wettest season is from January to April, the driest from May to August, and a transitional season from September to December. In addition, average annual rainfall and mean temperature are related to altitude. Increases in altitude are associated with decreases in both temperature and rainfall. Frost is a hazard beginning at about 2,100 meters (7,000 feet).

From the standpoint of both the stability of natural ecosystems and potential for human life support, the soils of Papua, New

Guinea, are poor. Soils are generally immature, moderately acid, and deficient in plant nutrients due to high rainfall-caused leaching. These problems are particularly prominent in montane zones. Such physical conditions produce vegetation with bulk but little protein and other nutrient content. This is reflected in the fact that many plants reproduce asexually, without producing flowers, fruits, and seeds, processes that otherwise involve the synthesis of amino acids. Hence herbivorus animals, particularly browsers, are relatively sparce.

Broadly speaking, there are three kinds of "natural" ecosystems whose characteristics influence the biology of man in montane New Guinea. The broad "major ecosystem" types considered are here designated *lower montane rain forest, montane forest,* and *montane grassland and savanna.*

Lower montane rain forest ranges from approximately 500 meters (1,500 feet) to 3,000 meters (10,000 feet) above sea level. The forests within this zone are evergreen, generally composed of broadleaf trees and/or pines, but they lack uniformity of composition between locations and altitudes. Two-tree strata and stands of single species are common; oak forests (lower part of the range), beech forests (middle altitudes), and pine forests (middle to high altitudes) occur along with the aforementioned mixed forests. *Pandanus* (screw pine) is particularly prominent in the upper portion of this zone along with tree fern, bamboo, and especially species of fig. The ground stratum consists of a variety of shrubs, ferns, and small creepers.

Lower montane rain forest gives way to *montane forest* (moss) at between 2,700 meters (9,000 feet) and 3,650 meters (12,000 feet) and at somewhat higher altitudes to alpine grasslands. The moss forest is continuously enveloped in fog, the ground is wet and spongy, and moisture drips constantly from the moss-covered vegetation. The lower trees are pines and species of myrtle and eucalyptus. Above the tree line the vegetation is a mixture of shrubs, ferns, tussocked grasses, and tall cane grasses in stands on boggy sites.

Strictly speaking, *montane grasslands* are not "natural" but are induced and maintained by human action on land that is potentially forest. They are included here as a significant type of ecosystem because they are widespread, enduring, and affect human life support. They are dominated by such grasses as *Imperata* and *Themeda* and are thought to result from human agricultural activity and the use of fire.

As a consequence of the controlling physical and biotic factors already described, New Guinea is one of the most impoverished areas of the world for vertebrate fauna. Hence the availability of mammalian fauna can be particularly important for interior and highland human populations removed from maritime and riverine zones where aquatic life might otherwise be exploited for high-quality protein. Within the limits described, the distribution and abundance of edible animals is related to vegetation and altitude. Forested areas are superior to grasslands, and lowland and lower montane communities up to 1,400-1,500 meters (4,600-5,000 feet) are superior to higher altitude communities. In addition New Guineans have for at least 6,500 years benefited from the introduction from Asia of the pig, both as a wild and a domesticated animal. The pig is, next to man, the largest edible creature on the island. Such factors bear upon the intensity of human exploitative patterns, particularly agricultural ones, to determine the local availability of wild animal foods.

Other sources of environmental problems for man can be cited briefly. One of these is the family of malaria parasites whose distribution is determined in part by the occurrence of the optimal habitat of anopheles mosquitoes, the vectors of malaria. This risk varies with altitude; malaria is endemic and epidemic in lower altitude zones. It has only become a problem at elevations above 1,500 meters (5,000 feet) with the increased human mobility fostered by colonial administration. Another problem endemic in the highlands is the limited availability of iodine in the environment and hence in the diet. Iodine deficiency may be related partially to the relative absence of meat in the diet; environmental iodine ordinarily would be concentrated in a series of trophic exchanges charcteristic of a long food chain.

The Distribution of Human Populations in the Highlands

The six broad intermontane valleys that are the centers of human occupation in the highlands have a profound demographic influence over their corresponding regions. There are, however, wide differences in population density among them. The highest densities in this series of centers are found in the Wahgi-Chimbu Valley with up to 230 persons/km^2 (600/mi^2) and the Lai-Enga Valley with 115 persons/km^2 (250/mi^2). The lowest in the series is Telefomin at 4 persons/km^2 (10/mi^2), immediately to the south of the Miyanmin. Figure 5.17 illustrates these population distributions.

The role of these valleys as centers of expansion in their regions is most likely based on the relative absence of extreme limiting factors, which has promoted a marginally faster rate of population growth.[36] This in turn has caused a ramifying pattern of expansion and out-migration in the direction of less densely settled areas, particularly lower altitudes. The lower-altitude areas, because of the operation of extreme limiting factors like malaria, have in turn functioned as population "sinks."

Life-Support Systems and Energy Flow

The illusion of uniformity apparent to many observers of man in the highlands of New Guinea is based on the existence of a common repertoire of plants and animals, both wild and domestic, together with the techniques employed to manage and exploit them.[37] All highland populations practice some form of agriculture dominated by one or more root crops such as taro, sweet potato, and yam, and also manage a variety of other garden and tree crops. The pig is universally important as either a wild or domesticated source of high-quality protein. Many highland populations also exploit a range of wild plant and animal species for their dietary value and for other purposes. Hence the flow of energy to man in the highlands is based on three or four basic food chains, the *root crop* chain, the *domestic pig* chain, the *"other crop"* chain, and the *wild fauna* chain. To these and their various manifestations should be added subsidiary chains relating various human populations to each other, and a chain involving the exploitation of *wild vegetable* products.[38]

The elements in the physical, or non-living, component of the ecosystem that con-

36. J. Stanhope, "Patterns of Fertility and Mortality in Rural New Guinea," *New Guinea Research Bulletin* 34 (1970):24-42.

37. M. J. Meggitt, *The Lineage System of the Mae Enga* (Edinburgh: Oliver and Boyd, 1965), pp. 267-8; A. P. Vayda, "Diversity and Uniformity in New Guinea," *Acta Ethnographica Academiae Hungaricae* 15 (1966).

38. Under this heading are included *nontrophic* transactions involving traditional or modern exchange, as well as the *trophic* exchange involved in cannibalism. The practice of cannibalism was once widespread in New Guinea and has nutritional importance. See, for example, M. D. Dornstreich and G. E. B. Morren, "Does New Guinea Cannibalism Have Nutritional Value?" *Human Ecology* 2 (1974: 1-12. *All* of these chains, where they occur, are particularly important for providing protein and fat.

trol the flow of energy in such systems include a range of limiting and tolerance factors; they determine the optimum environments for the elements of the living component of the ecosystem. Of particular importance are temperature (related to altitude) and rainfall. These factors have a principal role in determining natural vegetation zonation as well as the human utilization of suitable cultivated food plants. Of the root crop staples, taro demands the warm humid environment characteristic of a minimally transformed tropical rain forest (or other warm damp sites), the yam is a crop of warm dry regions and sites, and the sweet potato requires relatively dry conditions while tolerating lower temperatures (and higher altitudes).

Altitude, temperature, and rainfall influence the incidence of certain infectious diseases among human populations. As previously noted, malaria is inversely related to altitude and in some areas directly related to seasonal rainfall. In addition, the risk of upper respiratory infections increases with altitude. The availability of nutritional elements such as protein, fat, and iodine, which are ultimately attributable to soil conditions, may be limiting to man.

The internal loops controlling the flow of energy in these food chains consist of

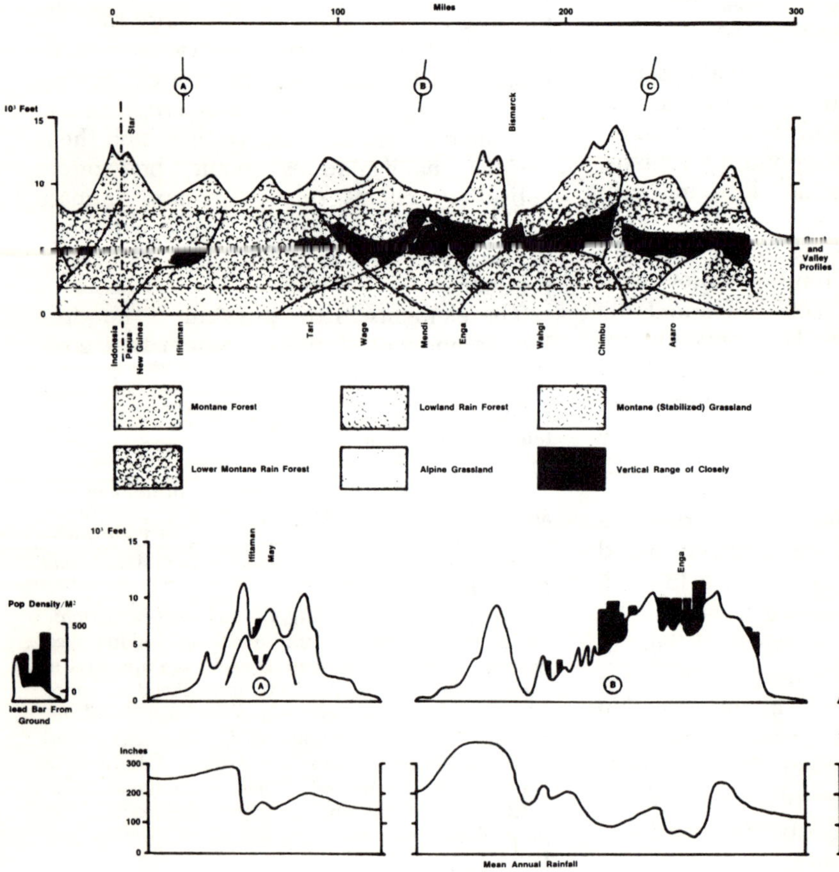

FIGURE 5.17. The distribution of a population in the highlands of Papua, New Guinea in relation to altitude, vegetation, crude population density, and rainfall. (Adapted from Brookfield, H.C. 1964. "The Ecology of Highland Settlement: Some Suggestions." *American Anthropologist* 66 (no. 4, part 2): 20-38.

the environmental effects of activities and movements of various species populations, including man, and are extremely complex and varied. Those related to the dominant controlling activities of man are not just confined to so-called subsistence practices, such as agriculture, hunting tactics, and the like. Rather, what is involved is a wide range of human activity which in particular life-support systems acts on work gates. Among other things these influence: (1) territorial integrity and acquisition, (2) the location of population clusters, (3) environmental modification, (4) the management of induced communities, (5) control of the primary production of certain plant species, (6) the degradation of natural communities, (7) the management of domesticate reproduction, (8) the movement of people, (9) the harvesting of plant and animal products, (10) transportation, (11) the political, ritual, or economic control of areally integrating chains, (12) the conversions of plant and animal foods into digestible forms, and (13) the control of factors directly influencing human biology (i.e., health, fertility, and rate of growth). To the extent that these kinds of human work gate inputs constitute a pattern of effects that maintain a balance between a human populaion and available trophic energy, this outline describes a *life-support system* as distinguished from the conventional *subsistence system.*

The spectrum of agricultural patterns found in the New Guinea highlands has been described in detail by the geographers Brookfield and Hart.[39] These authorities represent the spectrum as a continuum, with *land-extensive* swidden (cut and burn) farming patterns anchoring one end and *labor-intensive* forms of permanent farming anchoring the other end. Using the life-support system outlined above, a number of other continua can be associated with this spectrum, some of which have been explicitly recognized by Brookfield and Hart. These include: (1) increasing human population density, (2) greater settlement nucleation and permanence along with reduced movement, (3) greater degrees of land modification, (4) less use of wild fallow for soil restoration, (5) greater spatial differentiation of crops, (6) the expansion of domestic pig-keeping, (7) the expansion of areally integrating strategies, (8) the reduction of natural communities, and (9) the local exclusion of native fauna. Taken together these trends may represent a positive feedback system model describing the evolution of life-support systems in the New Guinea highlands.

The three tribal groups selected represent different points on this spectrum; the Miyanmin represent the extensive end, the Raiapu Enga the intensive end, and the Marings the intermediate. Energy-flow models for the three groups are presented; the systems are in roughly comparable scale and detail. Hence within the limitations imposed by the available data, we can draw for each group at least some comparable tween: (1) *energy production,* the food valconclusions about the relationships beue of plant and animal foods produced or made available by the system, (2) *energy consumption,* the value of the food consumed, and (3) *energy expenditure,* the amount of energy expended in human labor for life-supporting activities. The values for these variables have all been standardized in terms of their respective caloric values (in kilocalories) for the units of population involved, and man-hour equivalents have been provided for human labor. The caloric values can then be used to calculate energetic efficiencies (energy output per

39. H. C. Brookfield and D. Hart, *Melanesia: A Geographical Interpretation of an Island World* (London: Methuen, 1971).

unit of energy input) for the respective life-support systems and their parts.

The Miyanmin

The Miyanmin are shifting cultivators and hunters who live primarily in lower montane rain forest between 450 meters and 1,400 meters above sea level. The staple crop is taro, with a wide variety of other crops, such as sweet potatoes, bananas, leafy vegetables, sugarcane and the related pitpit cane, beans, and some pumpkins. Agricultural techniques are nonintensive. Forest clearance is only partial and of small scale, with a given plot cropped once before abandonment. There is no general burn of the site, no tillage, no attempt at drainage or erosion control, and rarely fencing. Tree crops such as pandanus, breadfruit, and sago are also cultivated. In season (the drier part of the year), wild plant foods, mostly the fruits of forest trees, are also exploited. Pig husbandry is very limited and casual. The reproduction of domestic sows is contingent on contacts with wild boars because all domestic boars are castrated. No extra gardens are kept to feed pigs, and no measurable work is attributable to "herding." Rather, pigs depend on wild forage, substandard taros, and "kitchen" scraps. The ratio of domestic pigs to humans is about 0.1:1. The most distinctive aspect of Miyanmin life support is the dependence on a success of the hunting of wild vertebrates, particularly mammals. Wild pig hunting is the most significant, but a variety of native marsupials and rats are also hunted, along with birds, snakes, lizards, and aquatic animals. Successful hunting is the product of work expended in settlement-pattern management and the movement of population between settlements as well as of direct hunting activity (see figure 5.18).

Before pacification in the late 1950s, warfare maintained territorial integrity in the face of expansive higher-altitude populations, permitted expansion against lower-altitude populations, and also resulted in the

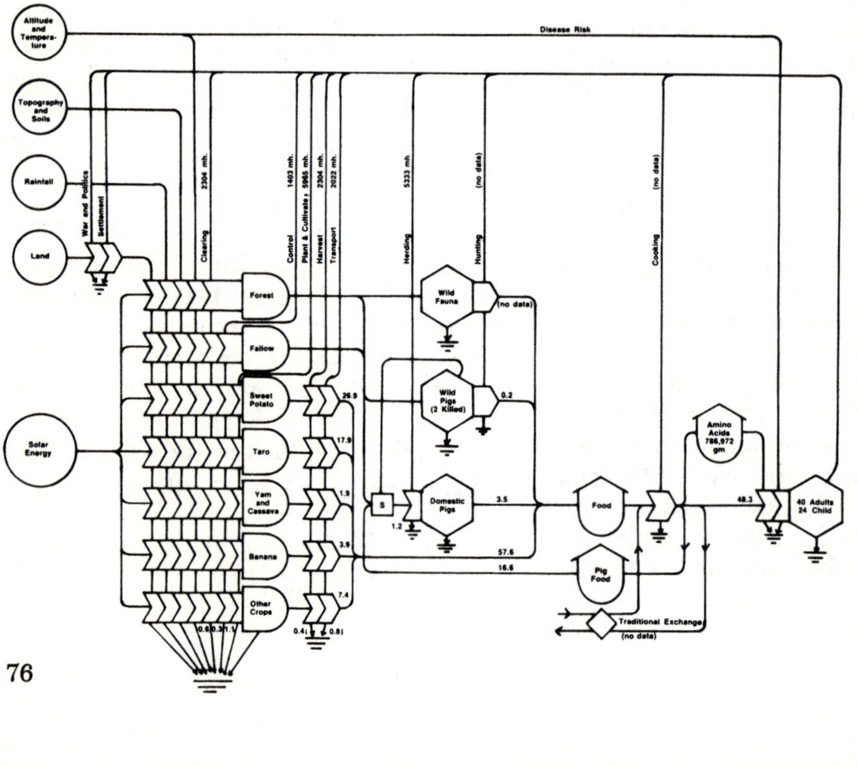

FIGURE 5.18. Miyanmin energy flow and control.

acquisition of human flesh and the capture of women and children (these latter are not included in figure 5.18).

The unit of population the model refers to is a residential hamlet, named Ukdabip, with a variable population averaging twenty-eight persons and centered around four households. This hamlet is one of eight such units that belong to a larger unit, a local population called Kome-Temsak, numbering one-hundred-eighty people. The larger unit is economically self-sufficient, whereas the constituent hamlets are constantly exchanging personnel and food among themselves. There are at least twelve such local populations spread out over as much as 5,000 km^2 (2,000 mi^2) in the upper Sepik region of New Guinea.

A variety of energetic efficiencies can be calculated for the overall model (figure 5.18) and its subcomponents. To deal first with a conventional view of the various subsistence strategies:

These calculations convey a simple-minded picture of the range of human activities which are in fact of vital importance. For example, in the foregoing discussion of the Miyanmin system it was pointed out that settlement-pattern management and movement of population contribute to sustaining yields in hunting. Similarly, energy expended on food preparation is absolutely necessary because most foods cannot be eaten raw, and because the availability of nutrients and digestibility are enhanced by appropriate preparation. Similar considerations apply to the "political" work gates and other kinds already enumerated. There are practical problems in implementing this suggestion because data on these nontechnological life-supporting activities are spotty. However, some first steps can be taken to further support the argument. The *net energetic efficiency* of the Miyanmin life-support system can be calculated as follows on page 78.

$$\text{Energetic efficiency of cultivation} = \frac{\text{food produced}}{\text{agricultural labor}} = \frac{(0.7 + 12.1) \times 10^6 \text{ kcal.}}{(1.2 + 0.5) \times 10^6 \text{ kcal.}} = 7.5$$

$$\text{Energetic efficiency of hunting and collecting} = \frac{\text{food produced}}{\text{hunting and collecting labor}} = \frac{(0.3 + 1.8 + 5.0) \times 10^6 \text{ kcal.}}{(0.04 + 0.06 + 0.7) \times 10^6 \text{ kcal.}} = 8.9$$

$$\text{Energetic efficiency of pig herding} = \frac{\text{food produced}}{\text{agricultural labor for fodder}} = \frac{3.5 \times 10^6 \text{ kcal}}{1.7\,\frac{3.9}{12.1} \times 10^6 \text{ kcal}} = 6.4$$

$$\text{Net energetic efficiency of subsistence} = \frac{\text{food produced} - \text{pig fodder} + \text{imports} - \text{exports}}{\text{subsistence labor}}$$

$$= \frac{(12.1 + 0.7 + 0.3 + 1.8 + 5.0 + 3.5 - 3.9 + 0.7 - 0.6) \times 10^6 \text{ kcal.}}{(1.2 + 0.04 + 0.06 + 0.5 + 0.7) \times 10^6 \text{ kcal.}}$$

$$= 7.8$$

$$\frac{\text{food produced} + \text{imports} - \text{exports} - \text{pig fodder}}{\text{all life-supporting labor}}$$

$$= \frac{(0.7 + 1.2 + 0.1 + 1.3 + 0.7 + 0.5 + 0.6) \times 10^6 \text{ kcal.}}{(12.1 + 0.7 + 0.3 + 1.8 + 5.0 + 3.5 - 3.9 + 0.7 - 0.6) \times 10^6 \text{ kcal.}} = 3.8$$

Another consideration in assessing the "success" of life-support systems relates to the role of essential amino acids (in protein) as a limiting factor in the diet of New Guinea populations. Limited protein availability may have a controlling influence on the flow of energy in systems involving man. There is evidence that New Guinea life-support systems "sacrifice" energetic efficiency in a strategic sense to meet the protein requirements of the population involved.[40] The Miyanmin life-support system provides an average of 65 g of protein per capita per day from an excellent balance of plant and animal foods. The proportions of this loss of efficiency will be apparent in succeeding examples.

The Tsembaga Marings

The Marings exploit a somewhat higher altitudinal range than the Miyanmin, but the bulk of their territory up to about 1,800 meters is within the lower montane rain forest zone. This is cultivable if topography permits, but the zone above this limit is honorable montane moss forest. There are tracts of savanna grassland in the Maring area that may be spreading slowly due to human activity. However, significant influences on Maring life support are not apparent, at least in the Tsembaga clan area to which the model refers (see figure 5.19). In the zone described the Marings practice a more intensive version of swidden agriculture than the Miyanmin. Gardens are completely fenced, work is expended on erosion prevention, gardens produce for several years in succession, and the ensuing fallow is controlled in terms of species composition. Sweet potato is the most productive crop, though more than half of it is fed to pigs, with taro being the single most important human staple. The Marings also grow yams, cassava, banana, beans, leafy greens, *Setaria* asparagus, sugarcane, and pitpit along with the recently introduced cultigens cucumber, maize, and pumpkin. There is still sufficient game in the area for productive hunting. The Marings also practice a moderately intensive form of pig husbandry with a ratio of pigs to humans that ranges from 0.8:1 before a major pig festival (involving extensive slaughter) to 0.3:1 after such an event. Like the Miyanmin, the Marings depend on wild boars to fertilize domestic sows.

Before pacification, Maring local populations were locked into a cycle of competition and warfare over territorial resources. This was undoubtedly connected with the larger pattern of demographic pressure and expansion in the highlands described earlier in this case study. Rappaport has provided a detailed systems model relating ritual to the regulation of this competition and of other ecological processes outlined here.[41] The model includes the dynamics of the pig herding and agricultural subsystems as well as the wide-ranging pattern of areally integrating trade and exchange. From Rappaport's discussion, it is clear that warfare and intergroup politics, and settlement-pattern mechanisms have important roles in the Maring life-support system. Unfortunately, this has not been quantified and can only be diagrammed in the present energy-flow model.

40. Refer to a discussion of this issue in chapter 3, particularly table 3.1.

41. Rappaport, *Pigs for the Ancestors.*

The model presented here refers to an "artificially" constructed unit of sixty-four people occupying one square mile of territory (see figure 5.19). This unit is taken to be representative of the Tsembaga, the "clan cluster" with a population of two-hundred-four, studied by Rappaport. This is just one of about fifteen Maring local populations living in the narrow Simbai and Jimi Valleys.

Within the limitations imposed by the available data, energetic efficiencies can be calculated for the Marings.

$$\text{Energetic efficiency of cultivation} = \frac{\text{food produced}}{\text{agricultural labor}}$$

$$= \frac{57.6 \times 10^6 \text{ kcal}}{(0.6 + 0.3 + 1.1 + 0.4 + 0.8) \times 10^6 \text{ kcal}} = 18.0$$

$$\text{Energetic efficiency of pig herding*} = \frac{\text{food produced}}{\text{labor for fodder + herding}}$$

$$= \frac{3.5 \times 10^6 \text{ kcal}}{[3.2 \frac{16.6}{57.6} + 1.2] \times 10^6 \text{ kcal}} = 1.5$$

$$\text{Net energetic efficiency of subsistence} = \frac{\text{food produced} - \text{pig fodder}}{\text{subsistence labor}}$$

$$= \frac{(57.6 + 3.5 + 0.2 - 16.6) \times 10^6 \text{ kcal}}{(0.6 + 0.3 + 1.1 + 0.4 + 0.8 + 1.2) \times 10^6 \text{ kcal}} = 10.2$$

Available data do not permit the calculation of life-support efficiencies in the Maring case. However, an examination of the efficiencies calculated above shows the energetic consequences of the strategy required to deal with protein problems. According to Rappaport's consumption data, the system is able to deliver an average of 34 g of protein per capita per day.

The Raiapu Enga

The Raiapu Enga exploit the altitudinal range between 1,500 meters and 2,000 meters above sea level in a grassland- and savanna-dominated intermontane valley. They practice a very intensive form of agriculture with continuously cultivated open fields devoted primarily to sweet potato. Less intensively cropped mixed gardens also produce sweet potatoes along with a wide variety of other subsidiary crops. The most important of these crops are yams, bananas, sugarcane, beans, pitpit, *Setaria* asparagus, and peanuts. Forest occurs only on ridge tops; hence hunting has little importance except as an adjunct to other activities (such as child play). On the other hand pig herding is intensive with as much as two-thirds of garden produce, primarily sweet potato and *Setaria,* going to feed pigs. The ratio of pigs to humans of 2.3:1 is reported as a maximum figure. Provision of a great deal of cultivated fodder for the pigs is necessitated by the almost complete absence of any sources of wild forage, either

* Compare this with: R. A. Rappaport, "The Flow of Energy in an Agricultural Society," *Scientific American* 224, No. 3 (1971):104-115 and M. Harris, *Culture, Man, and Nature* (New York: Thomas Y. Crowell, 1971) where a different procedure for calculating the efficiency of pig keeping was employed. See also table 3.1 in chapter 3 of this work.

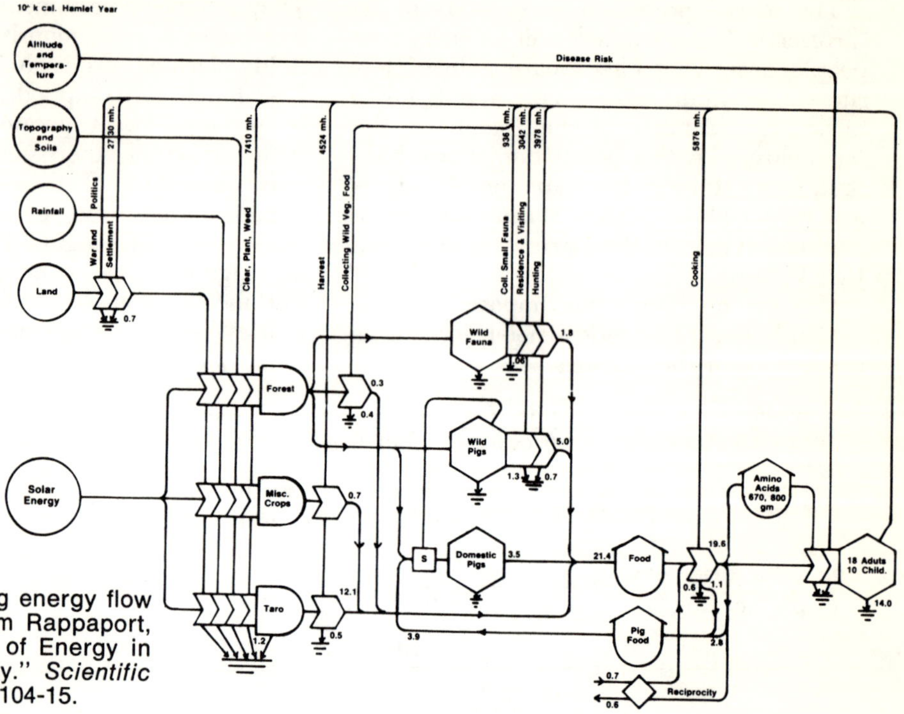

FIGURE 5.19. Maring energy flow and control. Data from Rappaport, R.A. 1971. "The Flow of Energy in an Agricultural Society." *Scientific American* 224 (no. 3): 104-15.

forest or river flat. Related to this is the local extinction of wild pigs, which requires the Raiapu Enga to keep domestic boars to impregnate their sows.

The intensive pig husbandry here is also related to the participation of the Raiapu Enga in one of the most extensive ritualized exchange cycles in New Guinea. This is the *Te* which has been analyzed from an ecological standpoint by Meggitt.[42] Reminiscent of the Maring situation, the *Te* functions in the political sphere to maintain clan boundaries in an area of intense competition and also promotes the movement of protein- and fat-rich food sources over an area exceeding 8,000 km^2 and including several hundred thousands of population. Food inputs and outputs to this *Te* system are not measured for the sample community, but ceremonial activity is quantified in the model (see figure 5.20).

In recent years the Raiapu have become involved in the modern cash economy (compare with the Altiplano Indian case study). This involvement includes growing special cash crops, such as coffee and Western vegetables, and buying canned goods, rice, and other goods from stores. The strategy is energetically insignificant, but it does provide 35 percent of the fat and 13 percent of the protein in the diet.

There is little shifting of houses and settlements in this system, though there is considerable individual movement and visiting in connection with life-supporting activities (not quantified here).

42. M. J. Meggitt, "System and Subsystem: The *Te* Exchange Cycle Among the Mae Enga," *Human Ecology* 1 (1972):111-24.

43. E. Waddel, *The Mound Builders* (Seattle: University of Washington Press, 1972).

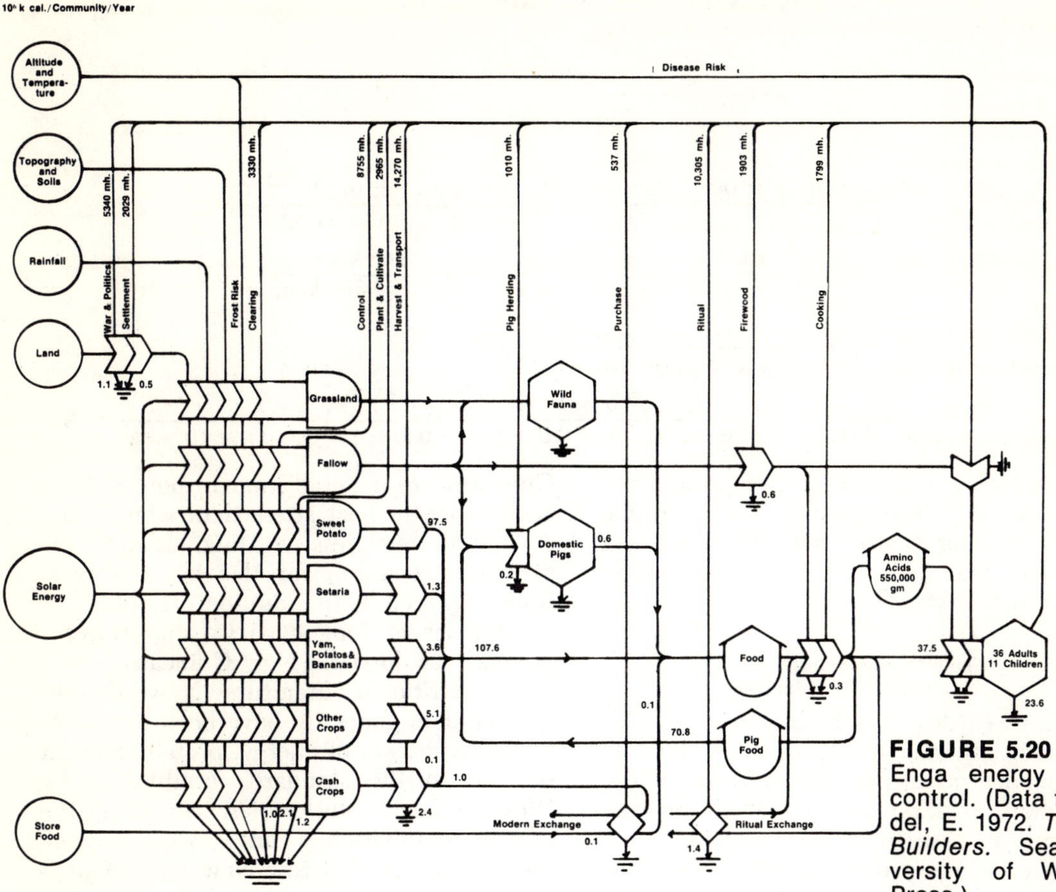

FIGURE 5.20 Raiapu Enga energy flow and control. (Data from Waddel, E. 1972. *The Mound Builders.* Seattle: University of Washington Press.)

The model (see figure 5.20) pertains to a sample community called Modopa chosen by Waddal to represent the larger local population.[43] It consists of ten households with a total population of forty-seven persons and occupies a section of natural terrace in the valley. Modopa is a small segment of Aruni clan, one of the one hundred seventy-six Raiapu Enga clans. The Raiapu in turn are one of two major localized cultural subdivisions of the Enga-speaking peoples who number about 140,000.

Energetic efficiencies, with a fair degree of comparability to those of the Miyanmin and Maring cases, can be calculated as shown below and on page 82.

According to the food consumption data provided by Waddel, this system makes

$$\text{Energetic efficiency of cultivation} = \frac{\text{food produced}}{\text{agricultural labor}}$$

$$= \frac{108.6 \times 10^6 \text{ kcal}}{(1.0 + 2.1 + 1.2 + 2.4) \times 10^6 \text{ kcal}} = 16.2$$

$$\text{Energetic efficiency of pig herding} = \frac{\text{food produced}}{\text{labor for fodder + herding}}$$

$$= \frac{0.6 \times 10^6 \text{ kcal}}{[6.7 \frac{70.8}{108.6} + 0.2] \times 10^6 \text{ kcal}} = 0.13^*$$

$$\text{Net energetic efficiency of subsistence} = \frac{\text{food produced} - \text{exports} + \text{imports} - \text{pig fodder}}{\text{agricultural work} + \text{herding} + \text{purchasing}}$$

$$= \frac{(108.2 - 1.0 + 1.0 - 70.8) \times 10^6 \text{ kcal}}{(1.0 + 2.1 + 1.2 + 2.4 + 0.2 + 0.1) \times 10^6 \text{ kcal}} = 5.3$$

$$\text{Energetic efficiency of life support} = \frac{\text{net food available}}{\text{total life-support work}}$$

$$= \frac{37.5 \times 10^6 \text{ kcal}}{(1.1 + 0.5 + 1.0 + 2.1 + 1.2 + 2.4 + 0.6 + 0.2 + 0.1 + 1.4 + 0.3) \times 10^6 \text{ kcal}} = 3.4$$

available an average of 32 g of protein per capita per day. This quantity is comparable to the Maring figure, but somewhat less than half that delivered by the Miyanmin life-support system.

Health and Biology of New Guinea Highlanders

It is useful to view the health status and biological characteristics of New Guinea Highlanders as in part the outcome of the operation of their particular life-support systems. In particular environments some systems are more successful than other systems in dealing with the variety of potential sources of biological stress to which human populations may be subjected. Accordingly, only a few generalizations are possible.

Health care is readily accessible to only that small proportion of the highland population situated near government and mission stations. In many areas of the highlands people may be within a day's walk of an aid post staffed by a native orderly who can administer first aid, basic antibiotic therapy, antimalarial drugs, and aspirin on a crisis basis.

New Guinea Highlanders appear to be physically fit with a well-developed musculature of the arms and legs in men and women. Dental health ranges from excellent in areas where taro is the staple to poor in sweet potato-dominated areas. New Guineans are generally much more able to acclimatize to heat stress than whites, and do so with much less sweating, although the precise mechanism for this has not been fully explored.[44] In addition, Highlanders display larger lung capacity and transfer factors than lowland New Guineans.

In nutritional literature New Guinea Highlanders generally have been noted for their low intakes of energy, protein, and fat relative to the standards established by Western health authorities. However, the standards themselves have been revised, and it is likely that few New Guinea populations fall appreciably below recommendations, at least on a group basis. As a matter of fact we have already shown how much effort New Guinea Highlanders expend in the energetically costly tasks of acquiring fat- and protein-rich foods. Of course, these efforts are not always completely or even partially successful. Some

*This value is low because, according to Waddel (see footnote 43), pig slaughters had been deferred due to organizational difficulties with the *Te* exchange system. This may also account for the relatively low level of protein consumption during the period of the study.

44. R. H. Fox et al., "A Study of Temperature Regulation in New Guinea People," *Philosophical Transactions of the Royal Society of London, (Series B) Biological Sciences* 268, no. 893 (1974):375-91.

areas of the highlands contain remarkably small inhabitants. In some cases individuals approach the size of and are mistaken for Oceanic Negritos. Such short stature is undoubtedly under a high degree of environmental control.[45] Among the groups described here, the Marings come closest to fitting this situation. In fact the Marings of the Simbai Valley display a range of responses to apparent protein and other nutritional deficiencies. According to Buchbinder, they display very slow growth and ultimate stunting, anemia, high goiter rates, low resistance to disease, late sexual maturation, and low fertility.[46] This is attributed to the synergistic interaction of protein malnutrition, malaria, and iodine deficiency, particularly acute among the highest altitude Maring groups (including the Tsembaga). Newly introduced stresses in combination with the factors cited have caused a long-term population decline in some Maring groups.

The Miyanmin lie at the other extreme. They are probably among the tallest and most robust people in montane New Guinea. Moreover, they appear to have recovered quickly from the low fertility and/or high infant mortality that occurred during the early contact period in the 1950s. Their population is expanding slowly despite the occurrence of malaria and goiter. The very high (by New Guinea highland standards) reported protein intake is certainly an important factor.

The Raiapu Enga population is also expanding slowly, and existing pressure on resources is aggravated by this expansion. Better health care in the area and the increasing incorporation of cash cropping, especially of coffee, are contributing factors in this population expansion. People with acute respiratory and gastrointestinal infections now have access to a good hospital. These two classes of infection have been a major cause of death in all age groups, with upper respiratory infections accounting for almost half of all infant deaths. Malnutrition does not appear to be as important a factor in Enga mortality as among the Marings, or among the Chimbu—most densely populated group in the highlands.

Yaws was once endemic in all parts of New Guinea, but it has been controlled very easily by antibiotic therapy. Malaria eradication programs have been successful in some local areas. Populations in the highlands continue to be susceptible to respiratory diseases, including chronic bronchitis, perhaps caused by "air pollution" from the warming nighttime fires. Pneumonia, usually preceded by influenza or a milder common cold-type viral infection, is the cause of moderate adult mortality in many areas. Gastrointestinal infections are endemic in high-population-density areas, like the Enga, and virtually absent in low-density areas, like Miyanmin. Chronic degenerative diseases, such as hypertension and arteriosclerosis, are uncommon among New Guineans adhering to a traditional life-style.

IN CONCLUSION

To study human populations within their respective ecological settings is an exciting and challenging endeavor. It is also an extremely difficult and complex task because of the inherent complexity of (1) the ecosystems of which humans are a part, and (2) the life-support systems that contribute to the maintenance and persistence of human populations. However, the rewards from such studies are great. By studying smaller, less complex human systems, we

45. D. C. Gajdusek, "Psychological Characteristics of Stone Age Man," *Engineering and Science* 33 (1970):26-52; W. W. Howells, *The Pacific Islanders* (New York: Charles Scribner and Sons, 1973), pp. 174-5.

46. G. Buchbinder, *Maring Microadaptation: A Study of Demographic, Nutritional, Genetic, and Phenotypic Variation in a Highland New Guinea Population,* Ph.D. diss. Columbia University, 1973), p. 213.

can not only expand our knowledge of the operation of more traditional societies, but we can begin to sort out the basic elements of human systems to predict the consequences of ecological interactions currently problematic in technologically sophisticated societies. Clearly as the complexity of the human life-support system and the dependence on outside resources increase, so too does the instability of the ecosystem. Attempts to manage imbalanced ecosystems by the imposition of new life-support systems or the input of higher levels of energy carries with it considerable risk. For these reasons studies must be made of relatively small and technologically simple human populations more closely integrated within their ecosystems.

The material presented in this brief exposition is actually based on very little ecological research. Our information on the ecology of Eskimos, savanna nomads, Andean Indians, and New Guinea Highlanders is still quite rudimentary. Fortunately, there are long-range plans to continue investigations of these and other populations within ecological frameworks. The ultimate goal of these studies is to be able to predict the time course of human-population dynamics, ecosystem changes, and the consequences of change in human life-support systems. At least with these predictive capabilities, we will be in a better position to make rational decisions about the future.

For Further Reading

Baker, P. T. and Little, M. A., eds. 1976. *Man in the Andes: A Multidisplinary Study of High Altitude Quechua.* Stroudsburg, Pa.: Dowden, Hutchinson, and Ross. A collection of twenty papers that summarize the IBP Andean Indian project.

Dyson-Hudson. 1966. *Karimojong Politics.* Oxford: Oxford University Press. An excellent study of an East African pastoral society with a focus on ecological perspectives.

Erickson, A. W., Hughes, D., and Milan, F. A. 1970. *Health and Biology of Circumpolar Human Populations. Arctic Anthropology* 7 (no. 1). A special issue devoted to reports from the IBP studies of Arctic Laplanders and Eskimos.

Rappaport, R. A. *Pigs for the Ancestors: Ritual in the Ecology of a New Guinea People.* New Haven, Conn.: Yale University Press, 1968. A classic work on the ecology of the Tsembaga Maring people from the New Guinea highlands.

Bibliography

Baker, P. T. 1969. "Human Adaptation to High Altitude." *Science* 142: 1149-56.

Brookfield, H. C. 1964. "The Ecology of Highland Settlement: Some Suggestions." *American Anthropologist* 66 (no. 4, part 2):20-38.

Clark, W. C. 1966. "From Extensive to Intensive Shifting Cultivation: A Succession from New Guinea." *Ethnology* 5:347-59.

Clark, W. C. 1971. *Place and People: An Ecology of a New Guinea Community.* Berkeley: University of California Press.

Dornstreich, M. D. 1973. *An Ecological Study of Gadio (Enga) Subsistence.* Ph.D. diss. New York: Columbia University.

Jamison, P. L. and Friedman, S. M., eds. 1973. *Energy Flow in Human Communities: Proceedings of a Workshop.* University Park, Pa.: IBP Human Adaptability Coordinating Office.

Morren, G. E. B. 1974. *Settlement Strategies and Hunting in a New Guinea Society.* Ph.D. diss. New York: Columbia University.

———. In press. "From Hunting to Herding: Pigs and the Control of Energy in Montane New Guinea" in *Human Ecology in the Pacific,* eds. T. Bayliss-Smith and R. Feachem. London: Academic Press.

Spencer, R. F. 1959. *The North Alaskan Eskimo: A Study in Ecology and Society.* Washington, D.C.: Smithsonian Institution, Bureau of American Ethnology.

Glossary

Abiotic Environment—The nonliving component of the environment, including the geology, topography, water resources, weather, and soils.

Adaptation—An adjustment or favorable response to some form of environmental stress by a population, individual, organ system, etc.

Altiplano—The high plateau of land that surrounds Lake Titicaca on the border of Peru and Bolivia. The base elevation of the altiplano is about 3,700 meters (12,000 feet).

Autotrophic—Organisms (usually plants) that are able to synthesize food from inorganic substances and sunlight.

Basal Metabolism—The minimal level of metabolic activity (in humans) that is compatable with life. Usually measured by determining the amount of oxygen that an individual consumes early in the morning, after a good night of rest, and in a fasting state.

Biomass—The weight of a group of organisms; either a population, community, or aggregate of populations.

Biome—A major ecosystem class, such as desert, coniferous forest, or grassland.

Biotic Environment—The living component of the environment.

Closed System—A system in which there is no inflow or outflow of energy, materials, people, etc. Most systems (e.g., ecosystems) are only closed in a relative sense.

Community—All of the living populations in a given area or ecosystem (often referred to as "biotic community").

Conductive Heating or Cooling of the Body—The transfer of heat to or from the body through direct contact with a solid object.

Consumers—That class of organisms that feeds on other living forms.

Convective Heating or Cooling of the Body—The transfer of heat to or from the body through convective movement of air or water that is in contact with the body.

Decomposers—That class of organisms, including fungi and bacteria, that breaks down, or decomposes, dead organic material into finer organic or inorganic constituents.

Demography—The study of the structure and dynamics of human populations, such as size, density, distribution, growth, and age and sex structure.

Ecosystem—An ecological system consisting of a biotic community of living populations and its abiotic component.

Ecosystem Diversity—A measure of the number of species and number of individuals per species in an ecosystem. A tropical rain forest is said to have high diversity; a desert, low diversity.

Endemic—A disease that is continually present in a given area (e.g., malaria).

Entropy—Randomness or diffuseness (of energy) within a system. The movement of energy from a concentrated to diffuse state is considered an increase in the entropy of the system.

Epidemic—A disease that enters a population, spreads rapidly, and then declines to low levels.

Evaporative Cooling of the Body—The transfer of heat from the body by the evaporation of sweat. Each gram of sweat evaporated removes about 0.58 kilocalorie of heat from the surface of the body.

Evapotranspiration—The evaporation of water from the surface of the ground, from bodies of water, and from the surfaces of plants. Transpiration refers to the evaporation of water from plant surfaces.

Fecundity—The biological or physiological capacity to reproduce (see also Fertility).

Feedback Mechanisms—Mechanisms of control, where a change in a given system contributes to another change in the same system. Negative feedback mechanisms help to maintain states of balance or homeostasis. Positive feedback mechanisms produce states of change that may then be modulated by negative feedback.

Fertility—Actual reproductive performance. In women fertility can be expressed as "total number of children born" or as "number of children born per year per woman aged fifteen to forty-five years" (see also Fecundity).

Heterotrophic—Organisms that depend on other organisms for food. They are unable to synthesize their own food as are "autotrophs."

Homeostasis—A dynamic steady state in a system that is usually maintained by negative feedback mechanisms.

Homeothermy—A form of homeostasis, where the body temperature is maintained within relatively narrow limits. Birds and mammals are homeothermic organisms. Reptiles, amphibians, and fish are "poikilothermic" organisms and conform more closely to fluctuations in environmental temperatures.

Hypoxia—A physiological state of oxygen deprivation in the body that is produced by the reduced oxygen content of air at high altitude. "Hypoxic" conditions exist at high altitude.

Infrared Radiation—Electromagnetic radiation that has a longer wavelength than visible, or light, radiation. Some infrared radiation is in the form of heat or is heat producing.

Kilocalorie (kcal)—A measure of heat and/or energy. The amount of heat necessary to raise the temperature of one liter or kilogram of water from 14.5°C to 15.5°C. The energetic value of food is usually measured in kilocalories.

Limiting Factor—Constituents of the environment that are available in minimal quantities with respect to a given population. A factor may be temporarily, seasonally, or permanently limiting (see also Tolerance Factor).

Montane—With reference to mountain areas.

Open System—A system in which there can be an inflow or outflow of energy, materials, people, etc.

Photosynthesis—The conversion to or synthesis of carbohydrate (sugar) from carbon dioxide, water, light energy, and certain catalysts (chlorophyll) in green plants.

Population—All of the organisms that constitute a given group or species (including humans) that inhabit a given area. Human populations are usually defined loosely and in a variety of ways.

Producers—Autotrophic organisms or plants that synthesize their own food.

Radiative Heating or Cooling of the Body—The transfer of heat to or from the body through direct movement of infrared radiation (see also Infrared Radiation).

Resource-Management Strategy—A set of practices that sustains a yield of energy or other important entities.

Respiration—The oxidation of sugar in organisms, where the energy is released in the form of heat and lost to the ecosystem.

Strain—The actual state of displacement or deviation from homeostasis in an individual organism or population. Strain is usually followed by some sort of restoring action (see also Stress).

Stress—A distorting action or force that arises from the environment that produces a deviation from homeostasis is an individual organism or population; also "stressor" (see also Strain).

Strategy Decision—A decision made by an individual or members of a human population that tends to facilitate the operation of their life-support system.

Subsistence System—One aspect of a life-support system that involves those activities associated with food acquisition or production.

Tolerance Factor—Constituents of the environment that are available in excessive quantities with respect to a given population. This may be a temporary, seasonal, or permanent state. The assumption is that a tolerance factor is damaging in some way to a population (see also Limiting Factor).

Trace Elements—Mineral elements such as zinc, iron, copper, and cobalt that are essential for a good diet but are needed only in very small amounts.

Trophic Pyramid—A pyramid representing the transfer of food energy (feeding patterns) from producers to successive consumer levels (see figure 2.1 for an example of a trophic pyramid).

Trophic Web—A network of feeding relationships within an ecosystem.

Ultraviolet Radiation—Electromagnetic radiation that has a shorter wavelength than visible, or light, radiation. Ultraviolet radiation can damage the skin through sunburn and is linked to the production of some forms of skin cancer. Ultraviolet radiation is also beneficial in that it facilitates production of vitamin D in the skin.

Ungulates—Animals having hoofs. Usually these are browsing and grazing forms of first-level consumers (herbivores).

Index

adaptation, 2, 6, 29
Allee, W. C., 2
anemia, 38
autotrophs: *See* producers

Baker, P. T., 32, 33, 42, 62, 66, 67, 69
basal metabolism, 30, 31, 35
Bates, M., 2
Black, R., 70
blood, as food, 56, 59
Blum, H., 11
body size, human, 31, 41-42, 52, 68, 83
  and surface area, 31, 60
  and temperature regulation, 31, 60
Boserup, E., 26
Brookfield, H., 70, 74, 75
Buchbinder, G., 83
Buffon, Comte de, 2
Bunzel, R., 22

Carlson, L. C., 30
Carneiro, R., 22
carrying capacity, 22, 42, 58
Centerwall, W. R., 41
cholo social class, 63
circadian rhythms, 33-35
circannual rhythms, 33-34
climate, 9
  and body size, 31, 60
  Andean altiplano, 62-63
  Arctic tundra, 46-47
  climatic stress, 29-35
  low oxygen pressure, 32-33, 63
  Papua New Guinea, 71
  rainfall (precipitation), 47, 54-55, 62, 71
  savanna, 54-55
  seasonal and daily variations, 33-34
  solar radiation, 31-32
  temperature, 29-31
cold stress, 30, 53, 68-69
Collins, P., 16
congenital malformations, 39
  carnivores, 8, 48, 65
  consumers: *See* heterotrophs
  herbivores, 8, 48, 65
culture, 4-6
  core, 4, 20
  defined, 6

Daniels, F., 31
Darwin, C., 2, 4
decomposers, 8
demographic stress, 42-44
dental health, 51, 67
Deshler, W. W., 54, 58
diet: *See* nutrition
disease, 39-42
  classification, 39
  defined, 39
  disease stress, 39
  east coast fever, 57
  endemic, 41
  epidemic, 40-41
  heart, 53, 68
  infectious, 40-42
    density-dependent factors, 40
    density-independent factors, 40
  kuru, 70
  malaria, 70, 72-73, 83
  respiratory, 51, 67, 68, 74
  soroche, 32
  trypanosomiasis, 51, 60-61
  tuberculosis, 51, 67
domestic animals
  alpacas, 64
  camels, 59
  cattle, 55-56, 60
  donkeys, 55
  goats, 55, 59
  horses, 64
  llamas, 64
  pigs, 72, 73, 76, 80
  sheep, 55, 59, 64
Dornstreich, M. D., 73
Draper, H. H., 52
dung, animal, 57, 66
Dunn, F. L., 41
Dyson-Hudson, N., 54, 55, 56, 58, 59
Dyson-Hudson, R., 54, 55, 56

East African savanna pastoralists: *See* human populations
ecological processes, 11
  control factors, 15
    homeostasis, 16, 29
    internal feedback, 15
    limiting, 15
    self-regulating, 15
    tolerance, 15
  energy flow, 11-14
ecology, 1
  and biology, 1
  and human sciences, 3-6
  and other sciences, 1
  cultural, 4, 5
  defined, 1, 2
  fundamentals, 7
  hierarchy of living things, 1
ecosystems, 2
  biotic community, 8
  components, 8
  diversity, 9, 41

Eskimos and tundra, 46
Indians and Andean altiplano, 62
in I.B.P., 3
New Guinea highlanders and tropical montane, 70
pastoralists and savanna, 54
productivity, 37, 55
terrestrial, 10-11
Eiseley, L., 2
energy, 11
conservation of, 14
consumption, 66, 75
degradation, 14
expenditure, 66, 75
flows, 12-13
Andean Indians, 65-67
Eskimo, 50-51
New Guinea highlanders, 73-82
savanna pastoralists, 58-60
structure of, 13
language, 12
laws of thermodynamics, 14-15
modules, 12-13
production, 66, 75
entropy, 15
environment
abiotic (physical, nonbiotic), 2
biotic, 2
environmentalism (environmental determinism), 4
environmental stress, 6, 29
Eskimos: *See* human populations
evapotranspiration, 15, 55

feedback mechanisms, 15, 29
Fenner, F., 40
fertility, 19, 42, 43, 44, 52, 83
Folk, G. E., 29
food, 7-8
chains, 14, 48, 73
production, 21
trophic pyramid, 8
Ford, J., 60
Fox, R. H., 82
Frisancho, A. R., 68

Gajdusek, D. C., 70, 83
gerontocide, 50
Godin, J., 49
growth and development, human, 52, 60, 68, 83

Hall, E. S., 48
Harris, M., 5, 20, 21, 79
Harrison, G. A., 70
Hayden, B., 42
health and biology
of altiplano Indians, 67-69
of Eskimos, 51-53
of New Guinea highlanders, 82-83
of savanna pastoralists, 60-61
heat stress, 30, 32, 35, 61, 82
Herbert, V., 38
herding: *See* livestock
heterotrophs: *See* consumers
Hiernaux, J., 60
Hipsley, E. H., 70
homeostasis, 16, 29
Hong, S. K., 34
human populations
African Bushmen, 41
African Pygmies, 41
Alakaluf, 30
Ama, 34
Arab, 38
Australian Aborigines, 41
Bantu, 61
Dinka, 60
Dodos, 54, 55, 58
East African savanna pastoralists, 53
East Asian Indian, 5
Eskimo, 8, 24, 30, 34, 44, 46
Karimojong, 53, 54
Maasai, 54, 55, 60
Malayan Semang, 41
Maori, 26
Mijanmin, 76-78
Montagnais-Naskapi, 5
New Guinea highlanders, 5, 22, 26, 69
Peruvian altiplano Indians, 25, 30, 42, 61
Raiapu Enga, 79-82
Tsembaga Marings, 78-79
Turkana, 53-54, 55
Tutsi, 60
United States, 21
Yanomama Indians, 40-41
human variability, 29
hunting, 48, 49, 50
Huntington, E., 4
Hurtado, A., 33
hypoxia, 32-33, 68

I.B.P. (International Biological Program), 3, 46, 61, 70
indigena social class, 63
infanticide, 50
iodine-deficiency diseases, 38, 72, 83

Jamison, P. L., 52
Kemp, W. B., 46, 48, 50, 51
kilocalorie, 11
Knobloch, H., 39
Kollias, J., 31
Kraut, H., 61
Kwashiorkor, 38, 42

Leeds, A., 22, 58
life-support systems, 18, 44, 49, 73
population control, 18
resource-management strategies, 18
structure, 18
limiting factors, 9, 15
Linnaeus, C., 2
Linton, R., 22
Little, M. A., 35, 53, 69

Maelzor, D. A., 2
Malthus, T., 2, 4
May, J. M., 39
Mayhall, J. T., 52
Mazess, R. B., 52, 69
Meggitt, M. J., 73, 80
mestizo social class, 63
Milan, F. A., 46, 52
milk, 58-59
Monge M. C., 32
Monie, I. W., 39
Moore, O. K., 5
Morren, G. E. B., 5, 26, 73
Murphy, R. F., 16

Neel, J. V., 41
Nelson, R. K., 47
New Guinea highlanders: *See* human populations
nutrition, 35-38
Andean Indians, 64, 67
and lactation, 35-36
and nutritional stress, 35-38
and pregnancy, 35-36
elements, 35
Eskimos, 52
New Guinea highlanders, 82-83
savanna pastoralists, 58-59

Odum, E. P., 1, 40
Odum, H. T., 12, 13, 58
oxygen pressure, low, 32-33, 62-63

pastoralists, 53, 61
Peruvian altiplano Indians: *See* human populations
photosynthesis, 7, 31
physical fitness, 52, 67, 82
Picón-Reátequi, E., 67

INDEX
Continued

Pimentel, D., 21
population: *See* demographic stress
possibilism, 8
Pratt, D. J., 54
producers, 7, 8, 48, 65: *See* autotrophs

radiation, 7, 31, 34, 62
  and melanin skin pigmentation, 32, 52
  and vitamin D production, 32, 37, 52
  infrared, 31
  ultraviolet, 31
Rappaport, R. A., 4, 5, 22, 70, 78, 79, 80
Rennie, D. W., 52
Roberts, D. F., 30, 31, 60
Robinson, S., 30

Sahlins, M., 22
Simpson, H. W., 34
skin pigmentation, 32, 52
slash and burn (swidden) agriculture, 22, 75
  labor-intensive, 75
  land-extensive, 75
Slobodkin, L., 16, 17, 18, 26
Smith, D., 22
soils, 10-11, 31, 57, 71-72
South Pacific Commission, 70
Spencer, H., 4
Stanhope, J., 73
Steinhart, C., 21
Steinhart, J., 21
Steward, J., 4, 20
Stini, W. A., 33
strain, 29
stress, 29
systems (open and closed), 7

Talbot, L. M., 55
Thomas, R. B., 63, 65, 66, 68
Thomson, A. M., 42, 61
tolerance factors, 15
Tustin, A., 16

Underwood, J. H., 38
ungulates, 55

Vayda, A. P., 4, 5, 16, 22, 26, 70, 73

Waddel, E., 80, 81
Wallace, A., 2
Watt, K. E. F., 40
Western, D., 55-57
Winterhalder, B., 63-66
work capacity: *See* physical fitness
Wyndham, C. H., 61
Wynne-Edwards, V. C., 16